ISBN 978-1-330-50237-2
PIBN 10070542

# Similar Books Are Available from
# www.forgottenbooks.com

INDISPENSABLE HANDY BOOKS.

# BRITISH FERNS AND MOSSES:

### DESCRIBING THEIR

## Haunts and Habits,

#### THEIR

## FORMS AND USES.

##### WITH

### NUMEROUS PICTORIAL REPRESENTATIONS.

Ye green ferns and flowers,
Belov'd in past hours,
Ere the young heart had yielded its gladness;
We gaze on you still
By the gush of the rill,
In the depth of our spirit's lone sadness.

LONDON:

WARD & LOCK, 158, FLEET STREET.

M DCCC LXI.

# PREFACE.

HAVING derived so much pleasure from the pursuit of Ferns and Mosses—so much gratification while roaming amidst the homes and haunts of these beautiful objects—I have been induced to contribute my experiences to the series of "Indispensable Handy Books."

Not all the elegancies of Potichimanie, Diaphany, Knitting, Netting, and Crochet, ever offered so much to the light fingers of the fair sex as a few cases of Ferns; and no amateur gardener ever found a fairer field for his enterprise than a Fern-bank, or a collection of Ferns in pots. A few years since none but enthusiastic Botanists paid any attention to them, and fewer still had attempted their domestication for purposes of study and ornament. Now we find them in windows, in gardens, and even in the dark front areas of town dwellings.

Still, Ferns are not to be grown with as much ease as geraniums; they are tender plants, and want ·careful culture and nursing, but they will soon repay their cost in glorious displays of beauty. But there is no mystery about them. The greater number of our native species can be made to·thrive and increase with a very moderate attention, provided it be of the right sort; and we shall here enumerate such particulars as will enable any person fond of Ferns and Mosses to maintain a respectable collection at but little cost, either of time or money.

# HANDY BOOK

OF

# BRITISH FERNS AND MOSSES.

---

## JANUARY.

WHY is it that folks care not for green mosses,
Except to pack their crates ?  Why do enthusiasts
Pore for sea-weeds until their eyes grow weak,
Beneath stern beetling crags, by rushing waves,
With no small peril, or of life or limb,
Yet pass unheeding by this lowly tribe,
The meekest of Earth's children ?

FERNS are not yet unfolded.  Our attention must therefore be directed to the brotherhood of mosses, those unassuming yet peerless members of a large community, growing in lanes and woods, or beside dripping rocks, uninjured
by wind, storm, or frost.

Look upon a bank of mosses when a sudden gleam of
sunshine lights up the wintry landscape, and a soft south
wind has caused a sudden thaw ! beautiful are they in their
freshness and luxuriance ; in their greenness too, for
mosses retain their verdure at all seasons.  A few ice-
spangles may gem their leaves, but these begin to melt, and
their most exquisite ramifications remain uninjured, even
by the heavy pressure of frozen snow.

Mosses were ever my delight ! they grew on the roots of
our old fantastic beech-trees, beside many a clear stream
that gushed from out some cavernous recess, and went

B

singing through the valley, in rocky lanes, and far up the precipitous sides of old stone quarries on the village common.

Oh! the modest beauty of those mosses! They rise before my mental view in all their variety and vividness— visions of by-gone days! I well remember every stone and rill, all brakes and glens, where best they grew; one deep lone vale especially, beside the old road that led from Stroud to Cheltenham, where Tradition lingers with her haunting tales, telling that fierce Danes and Saxons contended there in mortal fray—and hence its dolorous name of Dudcombe: but a lovelier spot the sun, perhaps, never shone upon; with its streams and trees, its sunny spaces covered with short herbage, and its wild rocky banks, where huge masses of lichen-dotted stones jutted forth among ferns and brambles.

Botanists speak much concerning the Poppy tribe. They point to the head while growing, and the rigid curvature in the upper part of the stem, which gives it a position impenetrable by rain or moisture. But not less wonderful is the formation of the common Hair-moss (*Polytrichum commune*), which grew profusely beside our streams, occasionally on the summit of high hills; for such plants as affect marshy localities abound also on places where clouds settle, from which they derive sufficient humidity for the purposes of vegetable life. It is found therefore far up the Lapland Alps, in company with others of its kind, on the verge of perpetual snow, and in the desolate wilderness of Lychselle Lapland, among woods or beside torrents, spreading like a carpet over dry open spaces, the resort of reindeer. Many a Lapland family, who perambulate from place to place during the short-lived summer, seek out spots abounding with the Great Hair-moss, and it forms for them both bed and bedding. They trace out a circumference with their knives, and readily separate the tangled roots from the meagre soil on which they grow; this done, they spread out the soft and elastic moss to dry in the hot sun, and having obtained a piece of the same size for a coverlid, they tranquilly lie down to rest amid those wondrous solitudes, where the sun, after a momentary setting in glory above the horizon, recommences his ceaseless course.

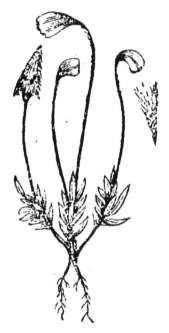

COMMON HAIR-MOSS.

The formation of the capsule is equally curious with that of the poppy. It is covered with an-elegantly shaped umbrella, such as Titania might carry in her hand; when the seeds ripen, the small cords that kept the umbrella in its place begin to loosen, some sportive zephyr jerks it in playful mood from off the capsule, and down it falls upon the earth. The stem then gradually bends, the straight stalk forms a curvature, and the seed-vessel being reversed, empties its seeds as from a pitcher.

Thus wonderfully is the moss constructed. Linnæus was, perhaps, the first who observed its conformation, when, journeying in quest of plants, he came into the savage wil-

derness where it overspread the ground in patches—the
haunts of bears, who slept soundly on this couch of Nature's
making, and of innumerable birds, who filled the woods
with melody, and chiefly built their nests with the finest
portions of this useful moss.

> ＼ "The farmer talks of grasses and of grain,
>  　　The sailor tells you stories of the main."

"It is therefore no wonder," says Linnæus, in his cele-
brated Oration at Upsal, when recurring to the dangers he
had passed through while exploring the wildest regions of
the North, " that I chose to make travelling in mine own
country the subject of my discourse. Every one thinks well
of what belongs to himself, and every one has pleasures
peculiar to himself. I have on foot passed over the frosty
mountains of Lapland in quest of plants; I have clambered
up the craggy ridges of Norland, and wandered amid its
almost impenetrable woods. I have made excursions into
the forests of Dalecarlia, the groves of Gothland, the heaths
of Smoland, and the trackless wastes of Scania. Truly
there is scarcely a part of Sweden which I have not crawled
through and examined, yet not without great fatigue of
mind and body. My journey to Lapland was an under-
taking of immense labour; but the love of truth and grati-
tude towards the Supreme Being constrains me to acknow-
ledge, that no sooner were my travels finished than a
pleasant oblivion of past suffering came upon me, and I was
richly rewarded by the inestimable advantages which I
gained from my labours."

Thus spoke the great Linnæus with respect to the benefit
which a man derives from travelling in his own country;
and his remarks may well apply to the pleasures that are
within the reach of all who seek to become acquainted with
the natural objects by which they are surrounded.

How many, ignorant concerning the ferns and mosses
that grow near them, complain that the country is dull in

winter! they hear of rich collections in Natural History, and are discontented with their lot; though many a rare plant, which men have brought with toil and hazard from far-off regions, is inferior to such as grow beside our paths at home. Go, then, into the woods and lanes, and collect those beautiful mosses, which strike their tiny roots into a thin soil, which root themselves in the crevices of large stones, or may be often seen through clear falling waters, that stream over the projecting rocks, beneath the arch of which they grow luxuriantly. I have gathered many such in days long past, and often look upon them with great pleasure, wondering that those who peril life and limb in obtaining sea-weeds, beside the roar of ocean, or among rock basins, or climb high cliffs in search of plants, should never think of collecting mosses. And yet mosses are beautiful and varied—the first, as regards their exquisite ramifications and often brilliant tints; the second, because no two of them are alike, though growing side by side, beneath the spray of the same fountain, or cherished by vapours ascending from the same damp soil.

Such plants have a prescribed use—not obvious, perhaps, but real. Like flowers, that successively emerge from out the earth, they each present a home, or storehouse, to some wayfaring insect; small birds select the finest kinds for their nests, and often feed upon the seeds that ripen abundantly at all seasons. The Bog-moss (*Jungermannia*), covers deep bogs with its spongy substance, and thus, by continual decay and renovation, produces abundance of vegetable mould, and turns them by degrees into fertile meadows. The loosely-matted patches are used for burning; and in many mining districts are in great request for forges.

The Long-stalked Earth-moss (*Musci phascum*) is a beacon-plant. If, in passing, its rich chesnut-red fruit-stalks and capsules are seen emerging from out a bed of verdure, you may fancy that a soft voice says to you,—

" This is the field of the sluggard ; it has lain untilled for
at least two years. 'Yet a little sleep, a little slumber, a
little folding of the hands to sleep : so shall his poverty .
come as one that travelleth, and his want as an armed
man.' " (Prov. vi. 10.)

" Of rills and fountains, guardian maid,"    .

the elegant and often-floating Greater Water-moss delights
in humid places.  The generic name, *Fontinalis*, designates
her station among fresh springs and rivulets ; the specific
one has a local meaning, yet, true to the former appellation,
this small moss is found on rocks and the roots of trees, in
brooks and rivulets, slow streams and ponds, but most especially beside cataracts, flourishing luxuriantly, where the
roar of headlong waters and their turmoil is greatest.
Hence the Water-moss thrives well in Sweden, and was
associated by Linnæus with many a legend-haunted spot.
The natives collect large quantities, with which to fill up
the spaces between their large chimneys and the walls of
their houses ; and thus, by excluding the air, to prevent
the action of fire.  The specific name was given in reference to this valuable quality.  Pale reddish tufts of the
*Conferva nana* may occasionally be seen attached to this
species in alpine rivers.

Those who feel strongly within them the "ardent inex-
tinguishable thirst of nature," which Cowper well describes,
and who, remembering the mosses they loved in childhood,
would again seek for such beside the hum of the great city,
may find the Lesser Water-moss (*F. minor*) growing abundantly on the walls of Lambeth Palace, fronting the
Thames.  Vainly, however, would they seek for the *F.
squamosa ;* of this the long and branched shoots uniformly
float in the direction of whatever stream it affects ; the
dark-green leaves become black when dry, and the whole
plant itself assumes such a glistening appearance that
Bauhin applied to it the epithet *lucens.*

Nor yet has the Feathered Water-moss (*F. pennata*), nor the Hair-like (*F. capillacea*), neither the Alpine (*F. alpina*), nor even the Lateral (*F. secunda*), been discovered in the neighbourhood of London, except in some old apple-orchard. The first, extremely rare, is assigned to the trunk of trees; it is mentioned by Mr. Drummond, in his "History of British Mosses," as growing on a beech-tree at Totherington, near Forfar; the second embellishes the stony banks of many an alpine rivulet in Scotland and Wales. Botanists may find it beside the gushing rills of the pass of Llanberris, within sight of Dolbaden's ruined tower; the third clusters, as already mentioned, on trees in orchards, and draws nearer to the vicinity of man than either the Hair-like or Alpine.

· The Leafy Buxbaumia (*B. foliosa*) must not be passed unnoticed. This plant, of frequent occurrence, especially in mountainous and open places, visited by the purest air of heaven, was named by Linnæus in honour of Dr. Buxbaum, a distinguished botanist, who sought out and enumerated such plants as grew around his native city Hol, in the Tyrol. Its relative, the Leafless Buxbaumia (*B. aphylla*), may he often discovered among fir-trees. This most singular of mosses can scarcely be said to have any stems; all that might be so called resemble a small bulb, covered with hair-like processes, but which, when highly magnified, are found to be true leaves, membranaceous, comparable to beautiful net-work, yet so narrow and minute as to be quite overlooked, or described merely as hairs. The whole moss is not an inch high; it presents a striking appearance when growing among others of the kind, being of a red colour.

The blood-coloured, or Obtuse-leaved Gland-moss (*Splachnum vasculosum*) is the finest and most beautiful of all British mosses; the fruit-stalk is one inch and a half high, upright and red, the receptacle large, spear-shaped, of a bright sanguineous hue, the fringe composed of eight minute

teeth, in pairs. Alpine tourists speak of it as growing in
bogs, and on the pointed masses of high storm-rifted mountains, such as Ben Lomond, and in extended patches as on
Ben Lawers, and the Clova mountains.

Another of the same family deserves brief mention. This
is the purple Bottle or Gland-moss (*S. ampullaceum*), which
localizes on bogs and marshes, often also on cow's manure;
growing about Ichen' Ferry near Southampton, at West
Wickham, Addington near Croydon, Geldestone Fen, Bun-
gay, and Suffolk. Ray mentions the second locality in his
" Synopsis of British Plants;" the association of his name
with a species still sought by modern botanists in the same
lone spot, invests it with no small interest. Nor less
attractive is it when found in a tributary stream north of
Tyfry, between that place and Hendref, in Anglesea, in
place where Druids dwelt, and where traces of their foot-
steps were found till lately, among circles of unhewn stones
and cromlechs; this place is mentioned by Mr. Davies as
well deserving the attention of all botanists, whether in
quest of plants or mosses, and also in each season of the
year. The veil is exquisitely bell-shaped, the receptacle
large, and resembling an inverted decanter with a convex
lid; and the fruit-stalks, two or three inches long, are
beautifully crimsoned. By attending to these characteristics the youngest botanist may identify the Purple Bottle-
moss, which ripens its capsules in July.

Others of the same family occupy spongy ground, or moist
places in alpine rocky districts, near Llyn Idwell, Carnarvonshire; or on Ben Lawers in the Highlands. Among
these, the Tongue-leaved Gland-moss (*S. lingulatum*), with
deeply indented leaves formed into cavities, and variable
fruit-stalk of a deep red, has attracted much attention,
both for its rarity and the difficulty of determining its
genus. It was first discovered in Scotland by Mr. Dixon,
and since on Ben Lawers and Ben Lomond by two brother
botanists; afterwards by Professor Hooker in muddy de-

clivities at the foot of Ben Cruachan, between Craigalleach and Meal-greadha, where it grew profusely, and afforded, in a fine August morning, a spectacle such as few muscologists have had the privilege of witnessing.

The beautiful Frælichian Gland-moss (*S. frælichianum*), with its fruit-stalk pale towards the summit, of a fine pink colour near the base, was first identified as a native moss by Mrs. Griffith, who found it on the eastern side of Snowdon, about two hundred yards from the highest elevation.

Look carefully for the Bryum-like Feather-moss (*Hypnum bryoides*). You may easily find it in shady places, woods, and ditch-banks, though very small; and this because of its capsules, which are edged at the mouth with a deep red fringe; the leaves are green, though not pellucid, and the reddish fruit-stalks issue nearly at the end of the shoots. The authors of " Systematic English Botany" have in their possession specimens of this little plant, gathered by the adventurous Mungo Park in the interior of Africa. He had preserved them with great care, and there is reason to believe that they formed the identical species to which he so feelingly referred when speaking of his own utter helplessness, and the powerful effect produced on his mind by observing the minute construction of this small plant. "I was very much cast down," he said, "and was beginning to despair, yet not without reason, for I was then in the midst of a wild country, ranged over by savage animals, and by men still more savage, five hundred miles from the nearest European settlement; and, considering my fate as certain, I was ready to lie down and die." At this moment the extraordinrry beauty of a small moss irresistibly caught his eye, and though unspeakably depressed, he could not look upon the delicate formation of its leaves and capsules without admiration. A train of soothing thoughts arose within him, and a consciousness that his Heavenly Father, who had thus called into being and preserved the tiny vegetable, beneath a burning sky,

and on an arid soil, would not desert an helpless traveller—
one whom He had made in His own image. Thus re-assured
the narrator went on his way, trusting that relief was at
hand, and he was not disappointed.

, Why is it that this small species, which affects our shady
woods and ditch-banks, should grow in Africa—that land
of cloudless skies and springless deserts? There are
problems in Natural History which the most learned cannot
solve.

· The blunt, fern-like Feather-moss (*Hypnum trichoma-
noides*), indigenous on the roots of trees, and in ditches
among woods, may be easily recognized by observing a
remarkable curvature in the scimitar-shaped leaf—a
peculiarity belonging exclusively to this species.

Few perhaps among our native tribes add more to the
picturesque effect of weather-beaten masses of rock or stone
than the *H. Halleri*, or Hallerian Feather-moss, discovered by Dr. Greville on Ben Lawers. This plant creeps
closely on its growing-place, in diffused tufts, of a rich
yellowish or reddish-brown colour, and is sometimes
pleasingly contrasted with the Waved Feather-moss
(*P. undulatum*), an exquisitely fine species, about a span
long, and of which the leaves are white and membranous.
The *Undulatum* mostly affects woods and shady places; it
is found also on Snowdon, and when closely examined,
exhibits a beautiful variety of tints in its component parts.
The fruit is long, slender, reddish; the veil straw-coloured,
with a brown spot at the end; rib of each shoot yellowish;
and the leaves tender, pellucid, smooth, shiny, and pale
green.

Nor less attractive is the reddish Shining Feather-moss
(*H. rufescens*), which thrives best where the torrent is
foaming. Its favourite locality is, therefore, the wet rocks
in the Highlands of Scotland, and nowhere is it more
abundant than on the perpendicular cliffs that start abruptly
by the falls of Moness.

The Squirrel-tailed Feather-moss (*H. sciuroides*) thus named from its numerous upright branches, simple and divided, and curving gracefully toward the points like that of the common squirrel, affords an interesting instance of restricted location. It is often found adhering to old trees; and though frequent in this country, is rare in Scotland, Intermoriston being its most northern habitation.

Years have passed away since I gathered from rock or stream side—from off the village common or old trees—specimens which have been treasured carefully, and still look beautiful. I recently opened them, and vividly did they bring before my mental view long-remembered scenes when life was new, and the future seemed as an unlimited horizon. I thought while looking at them of the pleasure which they had given me in their collection, and I could not help wishing that those whose attention during their summer and autumn rambles has been directed to sea-weeds and flowers, may go forth into the woods and lanes in this dull month, and derive from mosses equal instruction and delight.

## FEBRUARY.

" Mosses are Nature's children, no one careth
To make green merchandize of them; and yet
Nor sea-weeds, nor loved ferns, that quivering cast
Their shadows, or on rock, by rippling stream,
Or 'mid the wide heath, may compete with some
That I have gather'd."

Mosses are of considerable consequence in the vegetable world. The mould which they deposit rarely exceeds an inch in depth, and yet that small deposit is frequently all-important; their roots also, closely matted, and occasionally entangled with one another, serve to protect the rocks on which they grow from the effects of frost—from changes likewise consequent on the disengagement of carbonic acid gas—from fissures even in granite rocks, as in the neighbourhood of Clermont, in Auvergne. These changes, called by Dolomieu "la maladie du granite," resemble the dry-rot in wood, for the hardest blocks become soft, and readily crumble in the hand. Where, however, mosses assert their empire, the effect is neutralized by their absorption of the otherwise injurious carbonic acid, according to the well-known fact, that when two gasses of different specific gravities are brought into contact, even though the heaviest be the lowermost, they soon become uniformly diffused, by mutual absorption, through the whole space. By virtue of this law the heavy carbonic acid finds its way upwards in the lighter air of the atmosphere, and conveys nourishment to the tiniest moss or lichen which grows on the mountain-top.

In regions, on the contrary, where devastating torrents of rain fall suddenly, their transporting power is counteracted by a greater luxuriance of vegetation. A geologist, who carefully explored many parts of the tropical regions, observes, with reference to the fitness of all plants for the places they occupy, that the softer rocks would speedily be washed away in such portions of the globe, if the roots of parasitic and creeping plants were not so entwined as to present considerable resistance to the direct action of heavy rain. Unlike their forest brethren whose giant arms are vainly spread forth as if to repel the coming storm, and which are often hurled by its fury from stations in which they have stood for ages, the plants of which we speak serve to shoot off the fast-falling stream, and again lift up their heads when returning sunbeams light the dripping landscape. Mosses, in like manner, are not affected by running water, even if mountain springs assume in winter the character of copious floods, and rush impetuously over them.

The adaptation of plants to their respective positions, and the effects which their decay and renovation progressively occasion, are beautifully exemplified in the Bog-moss (*Sphagnum*).

This plant is fully developed in peaty swamps, and becomes, like the heath, a social plant ; or, in other words, it obtains exclusive possession of the ground, and lives in society. Such monopolies, however, are happily of rare occurrence, being checked by various causes ; for not only are many species endowed with equal powers of appropriating similar stations, but each plant, for reasons not yet fully ascertained, renders the soil where it has grown less fitted for the support of " other individuals of its own species, or even other species of the same family." Yet the tract, though occupied, it may be, by two or three usurping brotherhoods—who, to the exclusion of many others, are enabled throughout long periods to maintain their ground successfully against intruders, if even impoverishing it for themselves—is yet, by an irrefragable law of nature, improved for plants of another family. The tract thus appropriated may be an extensive moor, or a lofty mountain ; a sandy waste, or well-watered plain ; subject to equal diversity of soil or climate : still the operating causes which enable certain plants to maintain their ground against all others is equally developed, and the effects are everywhere the same. Oaks, for instance, render the sites whereon they grow more fertile for the fir tribe, and firs prepare the soil for the reception of acorns or sapling oaks, which thrive well. Every agriculturist, as Lyell justly observes in his " Principles of Geology," feels the force of this law of the organic world, and regulates accordingly the rotation of his crops.

The Bog-moss above mentioned, instead of deteriorating its place of growth, seems to have thrown a mantle over vast denuded tracts and unpeopled regions, preserving many a giant oak or pine that would otherwise have crumbled to dust; a way-mark, too, indicating traces of forgotten men, or implements of husbandry, and affording to the botanist and antiquary subjects of the deepest interest. This species cönstitues a considerable portion of all such peat as abounds in the marshes of Northern Europe. Peat may consist of any among the numerous plants that thrive best in moist situations, where the temperature is low, and vegetables decompose without putrefying—but the *Sphagnum* is by far the most abundant, in some portions nearly to the exclusion of all others, and possessing the singular property of throwing up new shoots in the upper part, while the extremities are decaying. Whenever, therefore, woods have been destroyed by fire—large trees uprooted by sudden storms of wind—or tracts of once cultivated land made desolate—embankments broken down, and marshes usurping the place of fertilizing streams—the Bog-moss rapidly takes root and flourishes.

In warm climates all decaying timber is presently removed by insects: termites and beetles with boring instruments set to work ; they perforate the wood in all directions, and when their ministry is accomplished, winds disperse the fragments to · incredible distances. It is otherwise in the cold temperature that prevails in our latitudes, and numerous examples are on record of the usurping powers of Bog-moss ; of its beneficial results also, and antiseptic property.

Thus, in Mars Forest, as related by Dr. Rennie, large trunks of patriarchal firs, which had fallen through extreme old age, were soon grown over by this friendly plant. We learn, also, that a sudden tornado having overthrown a considerable forest near Lochbroom, in Ross-shire, about the middle of the seventeenth century, its site was so quickly concealed from the same cause, that in about fifty years the inhabitants obtained peat. A similar instance is remembered with regard to the wood of Drumlanrig, in Dumfriesshire ; and old men tell their children, beside the winter peat-fire, that stately trees once grew where their fuel is now procured; they speak of the roaring winds and furious rains that drove against the old wood, and how, when warm in their beds, young children—as they then were—they trembled to hear the bellowing of the storm, and the crashing of the fine old trees that toppled down like ninepins one upon the other. The old men new nought concerning other peat-bogs, but the circumstance which they mentioned explains the occurrence—both in this country and on the continent—of mosses, wherein the trees were uniformly broken off, some close to the roots, others within two or three feet of the original surface, but all lying in the same direction.

In other instances, peat-bogs have originated from a different cause—the soil became, without doubt, too much exhausted for timber-trees, and, on the principle of that natural rotation which occurs in the vegetable world, one set of plants died out, and gave place to others. In proof of which, it has been ascertained that in the Danish islands, and in Jutland and Holstein, fir wood ·of various kinds (especially Scotch fir), is found at the bottom of peat-mosses, although it is certain that, during the last five centuries, no cone-producing plants have grown wild in those countries; trees of this family having been introduced towards the close of the last century.

We have mentioned, incidentally, the preserving ·quality of the peat, or Bog-moss; this quality is attributable to the carbonic and gallic acids, which issue from decayed wood, and is consequently absorbed by them; as also, to the presence of charred wood in the lowest strata of their vast accumulations; for charcoal is a powerful antiseptic, or preventive of corruption in animal and vegetable matter; and consequently, capable · of purifying water already putrid.

Nothing is more common than the finding of buried trees throughout the vast extent of peat-mosses. In those of Ireland, as also in most of such which abound in England, France, and Holland, they have been often observed with portions of their trunks standing erect, having their roots fixed to the subsoil, and consequently affording indubitable proofs that they once occupied the spot which now presents only a wild and denuded waste. In the marsh of Carragh comprising one of the wildest portions in the Isle of Man large trees are discovered, standing firm on their roots, though at the depth of eighteen or twenty feet below the surface. Indications, also, of large forests remain in Anglesea, beneath whose branches Druids reared their huts—the

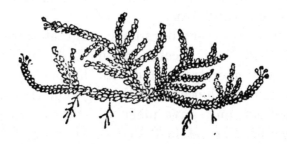

SPHAGNUM PALUSTRE—GREY BOG-MOSS.

very trees, it may be, around which waged the storm of war, when priestesses, with dishevelled hair, and torches in their hands, poured forth the most terrible execrations; and the islanders, stimulated to fury by their Druids, vainly sought to repel the troops of Suetonius. Be this as it may, the fact is certain—that Anglesea was one of the strongest holds of Druidism; and that her groves of oak, wherein human victims were often sacrificed, were cut down by command of the Roman general.

Some naturalists conjecture that trees may have been imbedded in peat-mosses through sudden eruptions of water; but the facts already mentioned show that such an hypothesis is inadmissible. It is likewise further disproved by the fact, that in Scotland, as in many parts of the continent, the largest trees are found in peat-mosses lying in the least elevated regions, and that the trees are proportionally smaller in such as occupy the higher levels. De Luc and Walker accordingly infer, that the trees grew on the spot, as timber uniformily attains a greater size in low and sheltered places. The leaves also, and fruits of several species, are continually found immersed in moss, together with the parent trees; as, for instance, leaves and acorns of the oak, the cones and leaves of fir, and nuts of the hazel.

It is more than probable, that no single plant throughout the vegetable world is so universally diffused as the Bog-moss. Other plants, doubtless, such as reeds and rushes, may be usually traced in peat, but wherever this substance is discovered, the *Sphagnum* constitutes its chief ingredient, and may be readily discriminated. When formed on a de-clivity in mountainous regions damp with springs, and where clouds continually rest, it scarcely ever exceeds four feet; when subsisting, on the contrary, in bogs and low grounds, it is occasionally forty feet thick, and upwards— which difference may, in some respects, be accounted for by the volume of water it contains.

And yet, though widely diffused, abounding in propor-tion to its distance from the Equator, and becoming not only more frequent, but more inflammable in northern latitudes —this valuable moss is subjected to certain laws, which re-strict its advance within the tropics. It is, moreover, rarely found even in the south of France and Spain; and although most plants contribute in warm climates to the production of peat, it is a singular fact, that neither the *Sphagnun*, nor any other kind of moss, enters into the composition of South American peat, which is chiefly formed of the *Astalia pumila*.

Our native moss is, therefore, never discovered in the Brazils; not even in the swampy portions of her vast allu-vial plains, drained by the sea-like Plata: on the eastern side of South America; nor in the island of Chiloe on the west. When, however, an English traveller reaches the

45th deg. of latitude, and botanically analyses the peat of
Terra del Fuego, the Falkland Islands, or the Chonos
Archipelago, he meets again the well-known *Sphagnum*,
which he has perhaps gathered in some green lane or wood-
side near his far-off home.

And yet, though locally restricted, a vast extent of
Europe is covered with this kind of moss. In Ireland, especially, it occupies—with different kinds of aquatic plants,
though in a far greater degree—a tenth of the whole island.
One of the bogs beside the Shannon is fifty miles in length
by two or three broad; and the great marsh of Montoire,
near the mouth of the river Loire—which gives a name to a
department of France—north of La Vendée, is more than
fifty leagues in circumference. It is also a curious and well-
authenticated fact, that several northern European mosses
occupy the place of pine and oak woods that have ceased to
exist within the historical era.

The same recent origin may be attributed to several in
this country. We have already instanced that of Lochbroom, in Rosshire; Hatfield Moss, in Yorkshire, may be
likewise mentioned. Local history preserves the fact, that
a vast forest occupied its site, about eighteen hundred years
since—a very ancient forest, without doubt, as prostrate
oaks have been discovered above one hundred feet long
fir-trees also, some more than ninety feet in length: all of
which were eagerly purchased for masts and keels of
ships.

The noble trees, which war or storms laid prostrate,
sheltered, without doubt, men of different races. Our
British ancestors dwelt among them; and recent drainage,
with the removal of peat accumulations, have laid open
Roman roads in the same moss of Hatfield, as also in that
of Kincardine, and several others ; a fact which, taken in
connection with the absence of British remains, goes far to
prove, that a considerable portion of the European peat-
bogs originated in the time of Julius Cæsar ; more especially

as the coins, arms, and axes, are clearly not more ancient than the era of his conquests. Nor can any vestige of the forests described by that general, and through which the great Roman road was formed, be discovered, except in the ruined trunks of trees, which the rapid growth of Bog-moss concealed for ages.

The aboriginal forests of Ardennes, Semama, and Hircinia, with others of equal extent, have long since disappeared, and their sites are occupied by swamps and mosses. That such vast sweeps of woodland once overshadowed a considerable part of France and Germany are facts pertaining to history; as, likewise, that their disappearance resulted from strict orders given by different Roman emperors to destroy both groves and woods throughout all conquered provinces. In after years, the same policy was adopted by Edward I. with regard to Wales; by Henry II. as respected Ireland. With the passing on of years, and the increase of civilization, different Parliaments made laws for the cutting down of extensive woods, because they harboured wolves and outlaws. No one, however, cared to remove a tenth of the prostrate trees; nor, indeed, could they, for trees were many, and labourers few. Wherever, therefore, the noble oaks came crashing down, there they remained; their trunks and branches obstructing the free drainage of atmospheric waters, and preventing many a bubbling stream that sprung from out the ground from flowing according to its wont. Mosses accordingly began their ministry; and brief space sufficed to enwrap, as with a mighty mantle of living green, those fallen fathers of the forest. Far as the eye could reach, even from the topmost bough of the stateliest oak or fir, would have appeared one wide, interminable mingling of forest trees;—now ascending some bold eminence, now stooping down into spacious valleys, then going on over ample plains, bounded only by the horizon. A few years passed: where the mighty had stood and fallen, was

seen only a wide extent of level or undulating ground, which
bore the name of moss.

Considerable tracts have, consequently, been reduced to
sterility as regards the growth of timber, by exterminating
edicts, and rendered less capable of administering to the
wants of man : but with the progress of civilization has
arisen a desire to appropriate them to purposes of agricul-
ture ; and hence, throughout many parts of England, bogs
have been drained : and rich fields of corn and homesteads.
reward the industry of the agriculturist.

Hatfield Moss, and that of Kincardine, with others of
great extent, bear witness to Roman triumphs, as already
mentioned.   Others still are, or were, belonging to a period
of unknown antiquity.   The body of a woman was discovered, about a hundred years since, in a Lincolnshire
peat-moor.   It was covered with moss about six feet deep,
and had lain there, apparently, for many ages.   The nails,
hair, and skin were scarcely, if at all, changed ; and the
antique sandals on her feet told of a widely-different condition of society.   It may be assumed that she was a person
of some rank—perhaps a British princess, or it might be a
female Druid—for sandals were confined to the higher
classes, or to those who ministered in idol services.   A
human body was likewise exhumed, a foot deep in gravel,
covered with eleven feet of Bog-moss.   It was completely
clothed in garments made of hair.   This curious circumstance occurred on an estate belonging to the Earl of Moira,
in Ireland : and the fact of hair garments identifies it with
a period antecedent to the one when British matrons learned
the use of the distaff from their German sisters.   No probable conjecture could be formed respecting the animal in
whose skin the ancient Briton had been enwrapped ; but
history leads to the conjecture, that the shaggy covering of
the goat was among the first materials employed in clothing ;
that afterwards the long hair of the caprine races was
blended with the short and soft fur of other animals, by the

aid of gum or glue, and manufactured into that coarse, but solid felt, known in Northern Asia from the earliest ages, and thus noticed by the poet:—

> " The careful pastor shears their hoary beards,
> And eases of their hair the loaded herds;
> Their camelots, warm in tents, the soldiers hold,
> And shield the shivering mariners from cold."

Goats'-hair, therefore, was the chief material used in ancient British vestments, till an improved condition of society led to the adoption, from Gaul, of the valuable arts of dressing wool, and of spinning and weaving cloth. Tradition tells, that such were brought into the island by a Belgic colony, about a century previous to the first invasion by the Romans. Authentic history relates, that an imperial manufactory of woollen cloth was established at Vinta Bulgarum, now Winchester.

Canoes, stone hatchets, and stone arrow-heads, evidently of British manufacture, have been found embedded in moss; as also skeletons of a gigantic elk.

Before dismissing this very interesting portion of our subject, we shall briefly refer to the origin of bog-iron ore, which is found occasionally at the bottom of peat-mosses. The frequency of this curious substance is familiar to the mineralogist, and its formation was long a matter of discussion, until the researches of Ehrenberg seem to have removed the difficulty. He observed, in the marshes about Berlin, a deep ochre-yellow or red substance, which, upon becoming dry after the water had subsided, closely resembled oxide of iron.

BOG-IRON ORE 2000 TIMES MAGNIFIED.

When submitted to a powerful microscope, the whole was

found to consist of slender, articulated threads, forming the cases of minute living creatures, called *Gaillonelle
ferruginea*. No doubt, therefore, now exists, that this
Bog-iron Ore—which is partly silicious and partly ferruginous—comprises millions of these small cases, which
although invisible to the naked eye, are yet so powerful in
their effects as to occasion the ebony blackness of such oaks
as have been found in peat.

For the sake of our readers who reside near peat-bogs, we
shall briefly mention, that the Grey Bog-moss is the most
common, with its two varieties :—the Zigzag is rare : the
leaves are of a splendid intense green, and when placed
under a water-spout, it assumes the character of a *bryum*.

Grey Bog-moss. Stems growing together, from three to
twelve inches high, upright : branches, two, three, or four,
from the same part—often drooping from the abundance of
moisture. Leaves white, egg or oval shaped, concave, soft,
tiling the branches. Capsules, when ripening on fruit-
stalks, urn-shaped ; generally several together at the top of
the stem.

Such are the natural history and associations connected
with the Bog-moss.

---

## MARCH.

" Where'er we search, the scene presents
　　Wonders to charm th' admiring sense,
　　　　And elevate the mind ;
　　Nor ever spreads a single spray,
　　That quivers in departing day,
　　Or turns to meet the morning ray,
　　　　But speaks a power Divine."

What is apparently more insignificant than moss-seeds ?—
in some species only to be discovered by aid of a high magnifier, in others resembling the finest grains of sand. What

is so utterly invisible as the wind—felt in its effects, but unseen? Yet the seeds and the wind, working conjointly, clothe the herbless rock with verdure, and form, as years pass on, a rooting-place for oaks that ride sea-billows, and circumnavigate the globe.

Mosses, therefore, and their handmaids, crustaceous lichens, are needful in the economy of nature: the first, as already noticed, prepares a slight accumulation of vegetable mould for the reception of the second; and these are rapidly succeeded by grasses and lesser plants, which in their turn decay, and give place to shrubs and trees, till after the lapse of years, extensive woodlands often clothe the boldest and most precipitous ascents. Thus, in the passes of the Alps, near Inspruck, the high cliffs on either side, though

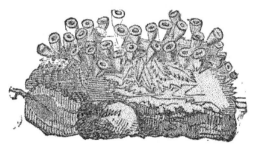

CUP, OR CHALICE-MOSS.

nearly perpendicular, are mantled with vast forests, that cast a dismal shade over the road. Time was, when those rocks were destitute of vegetation, when huge masses were raged over by fierce winds, and winter rains descended on them in their might: had the eye of some passing naturalist been open to discern things invisible, he might have seen a light vapour, borne by zephyrs, and left among the fissures of those wild rocks, where already the smallest particles of mould had accumulated; then came soft rains and sunbeams, ministering to the tender seeds, till forth from out their rocky cradles peeped green mosses of various forms and hues. The Cup, or Chalice-moss of old botanists, grew there abundantly, and its descendants still linger in

pen spaces beneath the trees; because, as wrote old
Gerard, "It thriveth best in moist barren' and gravelly
banks or rocks, creeping flat upon the ground, like unto
liverwort, but of a yellowish-white colour, among which
leaves start up here and there, certaine little things fashioned
like unto a tiny cup, called a beater, or chalice, and of the
same colour and substance of the lower leaves, which undoubtedly may be taken for the flowers. The powder of
this moss, given unto small children in any liqueur, for
certaine daies together, is a most certaine remidy against
that perilous malady called the chin-cough. Albeit, the
remedy doth require care, and is not to be adventured upon,
save under the guidance of an experienced gudewife."

The Toothed, hoary Thread-moss (*Bryum hypnoides*) is
found in the same locality. Concerning this, also, the old
herbalist has thus spoken :—"There is found, upon the top
of most barren mountains, but especially such as at whose
base sea coles are accustomed to be digged, or stones to
make iron of, and also where ore is gotten for tinne and
lead, a certaine small plant ; it riseth forth of the ground
with many bare and naked branches, dividing themselves at
the top into sundry knags, like the forked hornes of deere,
every part whereof is of a whitish colour."

The Northern Hair-moss (*Muscé septentrionale*) is there
also; that graceful species, first discovered on Ben Nevis,
and nowhere more abundant than on the highest of the
Cairngorm range of Grampian mountains, which thrives
best on rugged sides of windy rocks, where storms contend
for mastery. Yet, who, in looking on this moss, could
imagine that its delicate texture was adapted to bear the
merciless buffetings of winds and rains, unsheltered from
their fury, and covered half the year with a crushing weight
of snow. Yet so it is; and life is still sustained in this
small weed—a memorial plant, bidding him who looks
towards its sterile growing-place, take courage, wherever
his lot is cast.

Wherever a small stream wanders in the same wild loca-. lity, there the Rigid Thread-moss (*Bryum rigidum*) finds a home. Rills supplied by rain are not unfrequent on considerable elevations, and the naturalist who could ascend some of the most accessible, would often find the mosses that grow in valleys. This occurs in the instance of the Rigid Thread-moss, which thrives best where springs ooze from out the ground. Its diminutive relative, *Paludosum*, abounds in its vicinity, and may be seen on dripping rocks, or nestling among the massive roots of giant trees, which are rendered continually moist by extending in their neighbourhood.

Different species of the family of the Earth - moss (*Phascum*) are uniformly pioneers to their more attractive brethren. The Sharp-leaved Dwarf Earth-moss (*P. acaulon*) is extremely minute; but wherever it appears, with its soft and delicate leaves, a few lines in length, and forming globular clusters, he who passes away, and returns after the lapse of a few years, will find its herbless haunt covered with bushes, perhaps even with young sapling oaks or elms. The Beardless Earth-moss (*P. muticum*) may be readily distinguished by its red and yellow capsules, which become brown in autumn, and often present a pleasing contrast to the vivid green of various kinds. The whole plant is extremely minute; it attains occasionlly to an elevation of half an inch, though more generally is only three or four lines high. But however diminutive, the Capillary-branched Earth-moss (*P. serratum*) is still smaller. This fairy-formed moss resembles, at first sight, a thread-like byssus, and would be scarcely visible to the eye if it did not grow in patches. Conjectured to be a connecting link between the Musci and Algæ, partaking likewise of the nature of Phascum and Conferva, it consists of numerous filaments, which, when subjected to a magnifier, appear creeping, cylindrical, branched, and jointed like a conferva; the interstices pellucid, the joints darker green: and yet, how-

ever minute, and most probably the smallest of British
mosses, every part is elaborately adorned : the. egg-shaped
seed-vessels are pointed, and of a tawny hue when ripe ;
and the veil which serves to protect the seeds from the
effects of weather, or to hide them from the visitations
of small birds, is most exquisitely finished.

The *P. alternifolium*, or Alternate-leaved Earth-moss,
has its own specific character.  It forms small green tufts,
and the leaves, when examined separately, are short, awl-
shaped, alternate, rather bulging at the base, and expanding
at the ends.  The Crooked-stalked (*P. curvifolium*) in like
manner reveals specific differences, although hardly visible
to the naked eye unless growing in clusters, and bearing its
swollen capsules on small stems.  What, it may be asked,
are those peculiar differences?  Straightness in the spear-
shaped leaves that form the involucrum, or veil, while the
other leaves are egg spear-shaped, as also bending fruit-
stalks, terminated by oval seed-vessels, brown and mottled
when fully ripe.  Such are the peculiarities of this scarcely
visible moss, which render it different from any other of its
kind, as the yew is different from the poplar.  In the
Bearded Earth-moss (*P. piliferum*), we recognize a remarkable hoary appearance, occasioned by the long white filiform
extremities of the leaves.

The above-mentioned are most common among those
mosses which prepare the way for large vegetable developments, and enable seeds to germinate even in the fissures
and crannies of granite rocks.  In the Alpine passes all is
terrible and full of gloom.  Giant oaks, grasping with their
firm roots immense masses of overarching rocks, fling their
tortuous and rugged branches far over the defile, and often
reach the opposite bank, of which the summit is lost amid
the shade of intermingling boughs.

The beautiful vale of Tempè, on the contrary, offers an
instance of the fine effect produced by progressive vegetation.  Towards the lower part of this wild spot, the cliffs

are peaked in a very singular manner, and form projecting angles on the vast perpendicular masses of picturesque rocks that extend on either side the glen. According to the depth of mould, produced by the decay of lichens and mosses, are the fissures and ledges of the rocks varied with dwarf oaks, arbutus, and flowering shrubs.

Thus are we indebted to the gradual progress of vegetation in which mosses bear such a distinguishing part, for some imposing and many graceful varieties in scenery. Bare and rugged rocks may, in some situations, produce a grand, but never a beautiful effect; tinged with such concentric circles, nebulæ, and seeming pencilling of all hues and forms as lichens present, their sterile aspect disappears, mosses and ferns take root, and become objects of great interest to painters and botanists. To these succeed, or else mingle with them, flowers and small bushes—the dog-rose or honeysuckle, the daphne laurel, the dwarf cornel and mezereum. At this point, the rock acquires a considerable degree of beauty; but when clothed with forest trees, it becomes—especially if reflected by a sheet of water—one of the sublimest objects connected with natural scenery.

To such of our friends as live in the neighbourhood of those deep cuttings through rocks, which are made for the laying down of railroads, we recommend attention to this gradual advance of vegetation. It may not be that lichens and mosses first root themselves among the ruptured portions, because the rock being suddenly thrown open to the action of the elements, and affected by the escape of different gases, partially decomposes in many parts, and is consequently prepared for the reception of floating seeds. Progressive vegetation is, however, soon apparent, and becomes a subject of no ordinary interest.

The same effect may also be often traced on a common wall, and is equally deserving of notice. A green incrustation is first seen, composed of the earliest germination of some minute moss; when this decays, a very thin stratum of mould is deposited, which imperceptibly accumulates, and forms a soil for the reception of other mosses, and such diminutive plants as the *Drapa verna*, or Nailwort; others of a larger growth succeed, and before much time has passed, wallflowers, and the elegant snapdragon, with ferns and harebells, wave in the soft summer air.  The walls of the Jews' burying-ground, near the Queen's Elms, on the Fulham road, present a similar instance of vegetable development in its earliest and rudimental state.

Two subjects for consideration are suggested by the mention of mosses, taken in connection with trees or shrubs, and each has reference to that appendage which generally, though not invariably, acts as an anchor to the plant.  This appendage is the root; its fantastic form and tenacious grasp in various instances have been alluded to by poets and painters, and few, it may be, whose eyes are opened to admire the natural objects of creation, have not remarked the fine effect produced in broken foregrounds by the introduction of interlacing roots with ferns and mosses.  Salvator Rosa well knew the magic and truthfulness which they imparted, and many a painter since his time has visited the wildest solitudes of nature—green lanes and time-worn quarries, overgrown with old trees—in quest of such; poets have sung concerning them, and none more graphically than our own Spenser and the Bard of Avon.  Botanists, too, have loved to turn aside from the technicalities of science, to linger in imagination among forest walks, where moss-grown roots twist adown the banks, and are often embellished with primroses and bluebells.  Strange it seems, that among the sons of Painting or of Song, none have cared to find a theme for sketching or description in the elegant moss root, which binds its parent to some storm-beaten rock.  And yet the mystic apparatus of pipes and organs, of cells and vessels, equally exists in the minutest fibre of the Capillary Thread-moss—smallest of British mosses—as in the sturdy root that sustains the loftiest oak.  The machinery in each is similar;

the functions of absorption, assimilation, and secretion, with
the flowing of sap and the showing forth of its wondrous
powers, are the same in both. Moreover, I have often
thought, when endeavouring to remove a piece of moss, that
the power of adhesion in its roots is far greater than in
forest trees. Small though they be, and minute the green
patch which they sustain, they nevertheless strike their
sessile fibres so firmly into the rock or stone, that a sharp
penknife (and used by a strong hand) is often required to

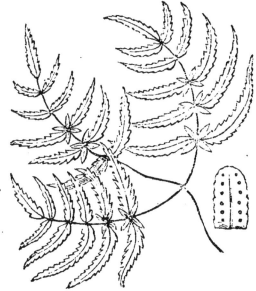

OAK FERN, OR WOOD FERN.

separate the moss or byssus from its place of growth. This
is needful; and were it otherwise, the tribe of which we
speak would be continually swept away by the mere force of
the wind or rain. Hence it is that the roots of mosses are
comprised under the general name of branching; several
kinds are furnished with small claspers, that possess great
muscular strength; others possess, if we mistake not, a
restricted power of adhering by means of suction.

Ferns are now beginning to unfold, and the botanist who

seeks for them in woods and bank-sides, may often discover round hairy-looking balls, of a rich brown colour, emerging from among the grass and mosses.   Such balls contain some infant fern, carefully folded up, but soon to yield to the joint ministry of showers and sunbeams, and to stand forth in its singleness and beauty.

Oak-fern grows generally in wild and mountainous districts, and, although one of the most elegant and attractive of our native species, seems instinctively to avoid the abodes of men, and fixes itself in places overhung with rocks or thick foliage.   The roots are black and fibrous, and the young fronds make their appearance in March and April; they each resemble three small balls upheld on wires, which gradually unfold, and display a triple division; the fronds arrive at maturity early in the summer, and entirely disappear before the storms of winter.

This species, the *Polypodium dryopteris* of botanists, derives its specific name from being occasionally found among the mossy roots of aged oaks.   Its localities are often associated with local scenery and time-haunted ruins, with the remembrance of Druidic observances and r tes, and places renowned in history.   Dry, stony heaths in Yorkshire, Lancashire, Westmoreland, and Scotland, are some of its favourite resorts, though growing in great luxuriance beside the fall of Lodore, on the side of Derwentwater, in Cumberland.   We have gathered it occasionally in Gloucestershire, in a wood north-east of the road up Frocester Hill, and on a rocky lane-bank leading to the romantic village of Shepscombe, near Painswick, towards the Cheltenham road.

The unfolding of this graceful species is ever welcome. Its emerging from the earth uniformly indicates the passing by of winter storms, and is accompanied by the lesser celandine, with its glossy yellow cups—the speedwell, and hawthorn, and those two most fragrant flowers, the violet and the meek, soft-eyed primrose. The mezereum, that fills the air with fragrance, and daffodils—

> " That come before the swallow dares, and tint
> The winds of March with beauty."

often affect the same locality.

BROAD FERN.

Fronds of the Broad-fern (*Lastræa dilatata*, or *Aspidium dilatatum* and *spinulosum*, and *Polypodium cristatum*, for by each of these names has the Broad-fern been designated) also appear in March, and although thus early developed, are rarely injured by the frost. New fronds succeed one another as the months pass on ; they apparently attain their maturity in September, and continue green and vigorous throughout the winter—yet only in sheltered places, for the Broad fern seems to shrink instinctively from cold.

This fern occasionally assumes a dissimilar appearance from such as it generally presents, and is therefore somewhat puzzling to inexperienced botanists. Four types are noticed by Newman in his interesting history of British species, and are as follow :—

The Linear type: erect, rigid, pale sickly green ; lateral margin of the frond nearly linear, as figured above.

The Dwarf type: dwarf, nearly erect, rigid, dark-green or brown ; lateral margin nearly linear ; all the divisions having a tendency to become convex.

The Triangular type: drooping, deep full green, broadly triangular ; the divisions slightly convex.

The Concave type: when luxuriant, drooping; when otherwise, more erect; triangular, bright beautiful green; all the divisions concave.

In every variety, the lateral veins are placed alternately on the mid-vein, after leaving which, each one sends out an anterior branch, which bears a nearly circular mass of thecæ half-way between its commencement and extremity. All the veins terminate before reaching the margin, which is attached on one side, but is soon lost among the growing thecæ, or sheaths.

The engraving represents the triangular or normal form, which gives a peculiar grace to this interesting species. Few among the brotherhood of Ferns are more widely-diffused throughout England, Wales, Scotland, and Ireland; growing not unfrequently on decayed trees, or on old stumps in hedgerows, on rocks and among stones; and is then, on account of its black, fibrous, and tenacious roots, extremely difficult to obtain entire; but when affecting woods and forests abounding with dead leaves, finding neither stones nor prostrate trees wherein to fix itself, the Broad fern may be readily removed.

---

## APRIL.

"Oh! I have loved where thou wast rear'd in greenest strength to stray,
And mark thy feathery stem upraised o'er licben'd ruin grey;
Or in the fairy moonlight bent, to meet the silvering hue;
Or glistening yet, when noon was high, with morn's unvanish'd dew."

FEW plants are more locally restricted than such as compose the Fern tribe; and yet this restriction cannot be ascribed to the want of shade or moisture. We remember a well-wooded park in Northamptonshire, watered by an ample stream, and having, moreover, a considerable morass, favourable to the growth of many species, where a few stunted specimens of the common Brake-fern (*Pteris aquilina*) alone were discoverable. They grew under the shade of trees, in somewhat swampy ground, and occupied a small space. Many, in passing, looked upon them with indifference; but their desolate and dwarfish appearance awoke within us the thought of fallen fortunes and stately homes exchanged for penury and obscurity.

COMMON BRAKE.

After leaving the growing-place of this isolated family, we sought carefully for more favourable specimens, but in vain. The hedges for many miles presented a rich luxuriance of wild roses and honeysuckles; and a beautiful variety of flowers common to the season was seen on either side the village roads; yet not a single fern. Nor was it till we reached Oundle, at sixteen miles' distance, that we observed small tufts of the Scaly Hart-tongue, springing from fissures in an old bridge which crossed the river Nen. One or two diminutive Polypodies were seen in the same locality; they

were, however, imperfectly developed, and had suffered
from the depredation of insects.

The absence of the Common Brake throughout such an
extended space is the more extraordinary, because there is
scarcely a heath or common—a wood or forest, in any part
of the United Kingdom, where it does not hold a prominent
station. Its presence is said to indicate a poor soil; but
Newman is inclined to think that its absence from rich and
highly-cultivated ground is rather attributable to the effects
of the plough and hoe. Varying in height from ten or
twelve inches to as many feet, it attains an enormous size in
shady woods where the soil is moist, and sunbeams rarely
enter. Kent is one of its favourite localities; and we remember gathering some fine specimens in a beech wood near
Ebworth Park, in Gloucestershire. The trees were large
and beautiful; but, with the exception of the Brakes and
common Solomon's Seal, nothing could flourish beneath their
shade : these plants, however, grew profusely; and it was
cheering to welcome them in a spot where even the common
green moss seemed disinclined to vegetate.

Young fronds of the Brake-fern appear in May; they are
very susceptible of cold, and the first shoots are almost invariably destroyed by the late frosts of spring, even when
the month is considerably advanced. They emerge from
out the earth either bent or doubled, the leafy portion being
pressed against the rachis; yet not curled, according to the
wont of other species. And as in spring this welcome fern
shrinks from such lingering frosts as seem unwilling to for-
sake the fields, so in autumn their visitations, however
transient, cause the leaves to become of a deep-brown colour,
and thus they continue during the whole winter, frequently
in an erect position, and affording shelter to small animals,
and birds when seeking for insect food.

The roots are brown, fibrous, and penetrating; the rhizoma
is also brown, velvety, of extensive and rapid growth, run-

ning mostly in a horizontal direction, though occasionally perpendicular. The historian of British Ferns, who watched with great interest the progress of the London and Croydon railway, found in the New Cross cutting, great abundance of rhizomata in a decayed condition, some of which had penetrated to a perpendicular depth of fifteen feet. And wherever this fern has grown unmolested for a long series of years, the soil becomes filled with a seeming network formed by them.

Seeds of the common Brake, equally with those of ferns in general, afford interesting objects for the microscope. The capsules in which they are contained, though appearing merely as dots or lines on the under-surface of the leaves, are either sitting or sessile, or else elevated on little foot-stalks, surrounded by an elastic or jointed ring, opening transversely when ripe, and dicharging the seeds—not merely causing them to fall upon the earth, but, by aid of the sudden jerk of the springing cord, flinging them to a considerable distance. During the months of September and October this curious mechanism effects its destined purpose, and sows a crop for the ensuing year.

FERN ROOT.

Many a schoolboy has wandered on a summer holiday from wood-side to sunny common, pleasing himself and his companions, as he passed along, with pulling up the finest Brakes and cutting their roots obliquely: wherefore? Because the roots, when cut, present a natural hieroglyphic, beautifully delineated, and representing either an oak tree or spread eagle. Some cavalier, it may be following the fortunes of Prince Charles through glen and glade where grew the Brake-fern in its wildest luxuriance, gave to this small painting by Nature's pencil the cognomen of "King Charles in the Oak." A mournful fancy, truly, had he who thus named it; but, ever since, the name has descended from sire to son, through generations of schoolboys, to the present day. Linnæus, in like manner, resting beside a rocky bank in Lapland, where the Brake-fern grew in such profusion as to form a canopy above his head, chancing to cut one of the stems a little way below the earth, found to his great surprise that it presented a kind of minute pencilling. Mindful of the Imperial Eagle, either as a cognizance of the House of Austria, or else having respect to the stern occupant of rugged mountains, winging his bold flight over regions of perpetual winter, he gave to it the name of *Pteris aquilina*.

The Pteris is not only abundant, but extremely useful; it is preferred in Scotland for thatching cottages and sheds, and serves in Wales for littering horses. You may see, even in the streets of London, cart-loads of this favourite fern at the doors of fruiterers and fishmongers. I often turn aside, when passing, to look upon its well-remembered branches; and many a thought arises of far-off scenes, where the Brake-fern flourished amid the loveliest haunts, by stream or wood-side, or on sunny heaths, among wild thyme and the bee-orchis.

Cottagers have recourse to the ashes of the common Brake for obtaining a tolerably pure alkali, mixed with water, and formed into balls, which are afterwards heated in the fire; they are much used to make lye for scouring linen. In countries where coal is scarce, the peasantry find them invaluable for heating ovens and burning limestone, for they yield a very great heat: when seen in the gloom of evening thus gleaming from some lone lime-kiln among rocks and aged trees, the effect is exceedingly pleasing.

A coarse kind of bread is prepared from the roots in some inhospitable regions of the globe; in countries, too, where fruits abound, and palms and citrons yield abundance of vegetable stores, the young shoots are often sold in bundles as a kind of salad. Those who prepare kid and chamois leather for sale employ ferns in dressing it; and often while the chamois hunter pursues his prey amid Alpine solitudes, his children range at the base of the stern rocks which he has ascended, in quest of this valuable fern.

And a truly wild plant is this same Brake, avoiding the haunts of man, and delighting in the purest air of heaven, among wastes and mountains, associated with legendary lore, and many a border tale of thrilling interest.

> " Beautiful fern!
> Thy place is not where art exults to raise the tended flower,
> By terraced walk, or deck'd parterre, or fenced and shelter'd bower;
> Nor where the straightly-levelled walks, of tangled boughs between,
> The sunbeam lights the velvet sward, and streams through alleys green.
> Thy dwelling is the desert heath, the wood, the haunted dell,
> And where the wild deer stoops to drink, beside the crystal well;
> And by the lake with trembling stars bestud, when earth is still,
> And midnight's melancholy pomp is on the distant hill."

TRUE MAIDENHAIR.

The True Maidenhair (*Adiantum capillus-Veneris* of authors) is the rarest and most beautiful of British ferns. He who adventures into moist caves, or on rocks near the sea-coast, may chance to find here and there a tuft of this elegant plant, firmly rooted in the crevices, yet uniformly

preferring a perpendicular surface, from whence its delicate
fronds spring forth in a nearly horizontal direction, inclining
upwards at the extremity.

In Cornwall, dripping rocks near St. Ives are favourite
growing-places of this rare fern ; as also a small cove on
the eastern side of Carrack Gladden, and a cove between St.
Ives and Hayle.   At the Lizard Point, that most southern
promontory of England, the scene of many a bitter parting
from those who are bound to the westward—botanists are
likewise rarely disappointed in collecting some of the finest
specimens.

Fern collectors who visit the principality of Wales during
their summer excursions, may find the species in a somewhat considerable range comprised within the rocks of
Dunraven, in Glamorganshire, and Barry Island.   A marked
restriction with regard to locality prevails in Ireland ;
although abounding with such dripping rocks as the Maidenhair principally affects, as yet it is discovered only in
the south isles of Arran ; among the Cahir Couree mountains, near Tralee, at the foot of a romantic rock facing
south-west ; and on the banks of Loch Bulard, near
Urrisbeg, Connemara.

Professor Beattie, who loved ferns, and sought them out
in their most secluded haunts, mentions the true Maidenhair as growing on the banks of the Carron, a rivulet in
Kincardineshire.

The generic name, *Adiantum*, is derived from two Greek
words, signifying *to moisten*, or *become wet*.   This elegant
plant is about five or six inches in height ; the leaflets are
fan-shaped, and of a very delicate transparent green ; the
roots are fibrous, black, and wiry ; the rhizoma black and
scaly ; young fronds appear early in May, though their
divisions are not fully developed before June.   They mostly
continue green till winter; but shrink instinctively from
storms and piercing winds : the botanist who then seeks for
them finds only a few dull brown branches, where a few

ays previously their graceful tufts looked green and cheerful in the fitful gleams of a waning sun.

We have mentioned the favourite localities of the True
Maidenhair; among these the south isles of Arran afford
ome of the finest specimens; and so abundantly grows this
most beautiful of our native species in their mild and hum₁
atmosphere, that the natives use a decoction of the fron s
instead of tea. They know not how eagerly fern collectors

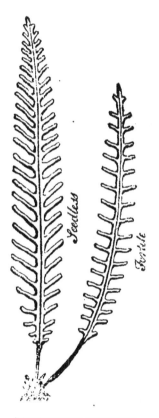

BLECHNUM SPICANT.

often adventure life and limb, scrambling up dripping rocks,
or exploring some lone sea-cave, in quest of the fern which
they scarcely heed; or with what delight the possessor of a
portable glass-house, when placing it in a staircase window,
or on some table in a favourite room, deposits within it a

young plant of the True Maidenhair, that he may watch its progress with the deepest interest.

The *Blechnum spicant* of Withering and Roth is now assigned to the genus *Lomaria*. Mr. Smith, of the Kew Botanic Gardens, restricts the genus Blechnum to those species in which the lateral, or side veins, continue beyond the line of thecæ, and to the margin of the pinna ; the genus Lomaria to such as present the lateral veins terminating in the line of thecæ. This distinction is extremely simple, and must steadily be borne in mind.

Few, if any, local associations pertain to this frequent species. It occurs on road-sides and village commons, in woods, by streamlets, and on moist heaths; in the southern counties sparingly, but more abundantly in the northern. The roots are wiry, black, and tough ; the rhizoma both tufted and hairy; emerging simultaneously from the earth with the lily of the valley, the cowslip and sweet violet, the white saxifrage and woodruff; it does not, however, again seek the shelter of maternal earth, but continnes green and luxuriant through the winter. The most casual observer may, perhaps, have noticed the beautiful arrangement of fern-seeds on the lower surface of the leaves; in some, profuse—in others, wanting. This peculiarity is very obvious in the *Lomaria spicant :* and some slight difference exists between such fronds or leaves as are called fertile, and such as are seedless ; it is not, however, sufficient to perplex the learner. Paley accounts for this singular arrangement of the seeds. " In all plants," said he, " two purposes are obvious ; viz., the perfecting and preserving the seeds." Seed-vessels are mostly lodged in the centre— the recesses or labyrinths of the flowers. They are surrounded with concave petals, which serve to reflect upon them both light and warmth ; and when any deviation occurs, it bears an especial reference to some peculiarity of flowering or station. Thus, in some water-plants, the perfecting of the seeds is carried on within the stem ; in the papilionaceous or pea tribe, a pent-house, formed of fragrant petals, protects the capsules from wind or rain. In the family of ferns, their seeds are placed either in spots or lines, and have, undoubtedly, regard to windy growing-places, on rocks or ruins, or open heaths where the species congregate.

Though neither historic nor poetic associations are awakened in the mind of him who gathers the long, slender fern-leaf—hear you not the voices of young children calling eagerly to one another? "Look! look!" say they, "what is that brownish green ball among the primroses?" and then, regardless of torn pinafores or wet feet, they scramble up the dripping or stony banks, among brambles and dog-roses, and in their eagerness too often spoil the desired prize. The ball itself is beautiful; and those who love to watch the gradual expanding of leaves and flowers observe with pleasure, that not only are the leaves rolled together, but the leaflets also. Remove one of the leaflets carefully, and you will discover on the back two white lines, extending from the base to the point, bordered with green, and depressed in the middle. These white lines are delicate membranes, containing minute pellucid bodies, supported on foot-stalks. High magnifiers, moreover, discover small bodies of a brownish cast on the youngest leaflets. They consist of two parts: the one, very slender and pellucid, proceeding from the rib; the other, a coloured oval-shaped ball, standing upon it. When the leaflets are fully unfolded, the rib becomes more turgid, and the globules disappear.

The Rock Brake, or Parsley-leaved fern, is found on rocks, and heaths, and old walls, especially in the northern counties. Tourists to Borrowdale, Cader Idris, and the Highlands of Scotland, may meet with this pleasing fern in many of their favourite haunts. It generally rises to the height of about four or five inches, and, when growing plentifully, its bright green leaves present a cheerful contrast to the lichen-dotted and dark weather-beaten masses of rock to which it clings. Though found occasionally in the cre-

vices of old stone walls, the Parsley-leaved fern thrives
best among the shapeless blocks of stone which time or
storms have strewn upon the sides of mountains.    In Eng-

SEEDLESS ROCK BRAKE.

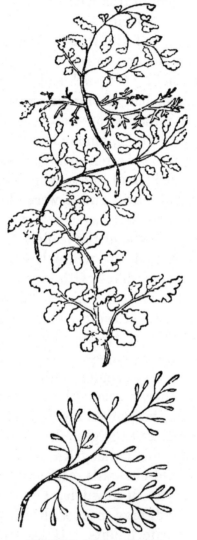

FERTILE ROCK BRAKE.

land, therefore, the lakes and mountains of Cumberland and
Westmoreland, of Lancashire and Yorkshire, reveal this
delicate species; the first two, very abundantly; the second,

more sparingly. A similar assignment and restriction is
discoverable throughout Wales. Botanists who visit the
mountainous regions of Carnarvonshire often meet with
specimens rooted among stones, which some convulsive
movement of the earth has shattered in times long past; let
them not, however, expect to find an equal abundance of
the Parsley-leaved fern on the sides or summit of Cader
Idris, or in the wild and beautiful localities of Dolgelly,
Aberglaslyn, Stranberris, or Beddgelert, with its rushing
waters, and rocks shaded with high trees, where, as tradition
says, the last of the Welsh bards used to wake the echoes
with wild and mournful melody; nor yet at Llanberris,
Tan-y-Bwlch, and Llyn Tregarien. Newman reports, that
he noted at least forty localities of the Parsley-fern during
the course of a pedestrian excursion in the Highlands, but
invariably in small tufts, on old walls, or among stones;
these localities occurred in the mountainous parts of Aber-
deenshire, Perthshire, and Argyleshire. In Ireland, the
Mourne mountains, County Down, and the liberties of Car
rickfergus, County Antrim, are mentioned as habitats of
the same fern, though sparingly distributed.

*Allosorus crispus* is the name assigned by Bernhardi,
Sprengel, Sadler, and Presl, to the Rock Brakes, or Parsley-
fern, the *Pteris crispa* of Smith and Withering. It has been
rendered the type of a new genus by three eminent botanists; Bernhardi gave it the appellation *Allosorus:* Desveux,
that of *Phorobolus;* Brown, that of *Cryptogramma ;* Linnæus called it *Asmunda crispa ;* Roth, *Onocleoides;* Gray,
*Stegania Onoclea crispa.* Young botanists will find it needful to remember these dissimilar names.

The root is fibrous; the fibres numerous and tough, and
tenaciously adhering to the wildest growing-places; hence
the Parsley-fern, though slight and delicately formed, is
enabled to retain its position on the side of mountains over
which the rains and storms of winter prevail unchecked.

The fronds, or leaves, appear in May, and disappear in

autumn, when frosts begin to whiten the fields. Fertile leaves, or such as produce seeds, are nearly triangular; they are composed of numerous separate pinnulæ, each on a distinct foot-stalk—the pinnæ, as well as the pinnulæ, being alternate. The character of the barren, or seedless frond, is various; it resembles in configuration that of parsley, being crowded, or crisped; but the divisions are intrinsically the same as those of the fertile, or seed-producing frond: in both the rachis, or spike-stalk, is slender, smooth, pale-green, and naked for rather more than half its length; the colour of the frond is of a bright and beautiful green.

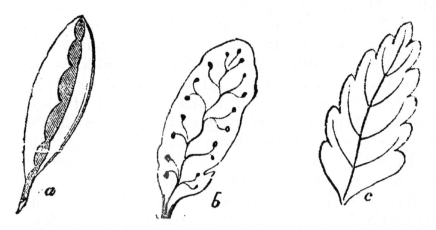

Fig. *a* represents a seed-producing frond, when the margins are rolled over in order to protect the thecæ. Fig. *b*, a leaf showing alternate lateral veins, which are generally forked; with a mass of thecæ attached at each extremity. The veins do not reach the margin. Fig. *c*, a seedless frond.

# MAY.

" MANY a poet, in his lay,
  Told me May would come again ;
  Truly sang the bards, for May
  Yesterday began to reign !
  She is like a bounteous lord,
  Gold enough she gives to me—
  Gold ! ay, such as poets hoard,
  ' Florins of the mead and tree ;
  Hazel flowers, and fleurs-de-lis !'
  Ferns that grow the stream beside,
  Where the leveret loves to hide."

  DAVYTH AP GWILYM.

AWAY to the woodlands ! to the mossy bank, and stream-side, in quest of ferns—to the rock or wall, the wild heath or sunny dingle—there grow these loneliest children of the spring or summer ! and scarcely may the wind or shower fertilize the dreariest crag, or a wandering sunbeam visit the most secluded cavern, where you cannot find them.

Take, therefore, a small basket, or tin case, and collect such specimens as you desire to preserve ; it may be that you have merely a space of a few yards, yet this, with judicious care, will become a Fernery. Observe, when gathering your specimens, the situations in which they grow—whether on an horizontal or sloping surface ; whether rooted in the ground, or simply adhering to some weather-beaten rock ; whether exposed to storms or sunshine ; and, according to their respective growing-places. arrange them in your Fernery at home. If you live in the neighbourhood of a glass-house you can obtain abundance of clinkers : if not, in this building age you will have no difficulty in procuring pieces of broken bricks, with which to imitate the ruggedness of nature. Fill some of the interstices with crumbling mortar, for the reception of those ferns that grow naturally in the crevices of mortared walls ; and they will, despite of

rains and constant waterings, in which ferns delight, remain comparatively dry: this is needful, because, although
the species mostly abound in humid places, some are injured
by too much wet. Bog earth, or leaf mould, will afford an
excellent rooting for moisture-loving ferns. Possibly neighbouring trees, high walls, or tall unsightly buildings, may
shut out a summer's morning sun, or even not permit a
single ray to illumine some dark corner. This, however,
need not perplex you: the corner, cheerless though it be,
and necessarily damp, will afford a welcome habitat to the
*Scolopendrium vulgare,* or Hart's-tongue, which especially
delights in old wells and humid places, and is nowhere so
abundant as in deep shade and moisture. Notwithstanding
these apparent predilections, it will be well to place ferns of
dissimilar localities side by side, in the deepest shadow and
brightest sunbeams; such, for example, as the common
Hart's-tongue, and its relative, the Scaly. You will then
be able readily to observe how luxuriantly the one expands
and seemingly rejoices, either in shade or sunbeams, while
the other appears to pine for a more congenial habitation.

It is all-important that ferns should be well watered, and
yet as gently as possible. If you possess a garden engine,
let the stream descend in an almost imperceptible shower;
if you have only a watering-pot, hold it high, and avoid a
heavy watering. Nature teaches this: for rain rarely in-
jures by its force even the feeblest flower. If the day has
been cloudless, refresh your ferns, and that copiously,
every evening during summer. In autumn withhold your
hand; such as conceal themselves beneath the earth in
winter begin to prepare for their long sleep; in others, the
fronds have ceased to grow—while some that remain green
and render cheerful many a leafless hedge or rocky bank,
have already their full size and substance developed. But
however circumstanced, they all require perfect rest; the
sap scarcely circulates—a state of vegetable quietness ensues,
and they cannot be disturbed or stimulated without injury.

Mosses may be introduced with great advantage; they speedily cover the earth or stones, and retain humidity from dews or showers; and if you wish to please your children, plant among them a few primroses and harebells. The sight of them may even recall to your own mind the gladsome days of childhood, when you gathered such among the grass, or beside some babbling stream rushing from out a wooded bank. The snowdrop, and a tuft or two of cowslips, will

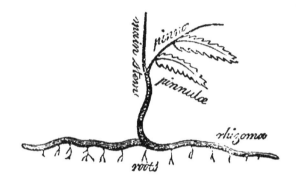

l.ok beautiful amid the ferns; it may be that, when in spring they lift up their familiar faces, you will incline to give your wife and children a treat into the country, far away from the sight of crowding houses. Such holidays refresh the spirit; they fill the mind with gladsome thoughts and pleasant memories, and he who occasionally enjoys them returns invigorated to his daily duties.

The owner of a small court cannot do better than embellish it with ferns. They grow where flowers yield no beauty; and, instead of that unsightly and desolate aspect which courts often present, a beautiful assemblage of graceful plants may be readily brought together.

While recommending the study o these interesting plants to those especially whose visits to their growing-places are few, and somewhat restricted, we desire to impress upon their memories the component parts of every fern, viz.—the *roots*, *rhizoma*, and *fronds*. The small fibres above pictured represent the roots: the long tube-like horizontal stem is part of the rhizoma, properly called creeping, because it extends a long way beneath the soil : but when otherwise, is known as a tufted rhizoma ; the upright stems are fronds, by some botanists designated branches, by others leaves. This part comprises a main stem, which extends from the rhizoma to the extreme point, and is called the *rachis:* the branches on either side are called *pinnæ;* when not completely divided from each other, as in the Hard-fern, *pinnatifid;* if divided, *pinnate.* When the pinnæ are divided into branches on both sides, the branches are called *pinnulæ;* of this the Brake presents a familiar example, as also of a further division into *lobes.* *Thecæ,* when applied to ferns, signifies capsules, or small vessels for containing seeds; these are beautifully arranged, on the under-surface of the leaves, in dots or lines. In those pertaining to the Marsh-fern, a small white kidney-shaped spot is obvious, consisting of a membranous substance, called the *indusium.*

But many an enthusiastic lover of ferns and flowers has not even a small court to call his own. He occupies, perchance, a single room, and sighs in vain for the possession of those beauteous ferns which delighted him in youth. We will cheerfully point out a simple method by which his wish may be fulfilled. Obtain from the glazier four pieces of glass, equal in size and thickness—from the linen-draper a piece of scarlet galloon (for this colour suits well with green), and bind it tight round the edges ; fastening it at the ends firmly with a needle and thread of the same colour. This done, sew the edges together, and form a square glass frame, to which a cover must be fixed by the same means ; provide a thick square board, with a groove all round, the size of your glass-house, or, what would be preferable, a strong box, sufficiently deep to contain five or six inches of light sandy earth ; plant in this a few young ferns of different kinds ; moisten the earth slightly, and cover them with your glass-house. You may thus have a continual object of interest to greet your first awaking: ferns, it may be, that grew beside your father's cottage, or where you gathered nuts in autumn, associated with thoughts of home and boyhood pleasures, bringing back to mind many a word of loving counsel to guide and cheer your onward progress.

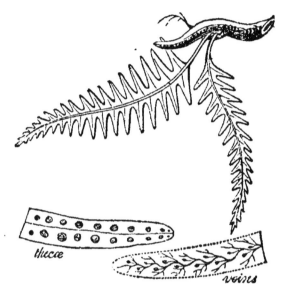

COMMON POLYPODY.

Lift up your head, young botanist, and think not that ferns grow only on the ground! A beauteous brotherhood of the Common Polypody *Polypodium vulgare*) is looking down upon you from the summit of a beetling crag; and yonder old pollard is crowned with a tuft, among which the carlet-leaved crane's-bill and blue-bells are waving lightly in the breeze of summer, and a little linnet is pouring forth his melody. There are many happy creatures among that tuft of Polypodies. See you not the sulphur-coloured butterfly, and her sister with gorgeously tinted wings—a few industrious bees, singing at their work, and the emerald-coated beetle taking a nap among the lichens? We will not, however, speak of these, but rather of ferns and the common Polypody—one of the best known and most abun-

E

dant. A friendly fern is this same Polypody—found in almost every hedge beside our paths, on the surface of storm-beaten rocks and deserted ruins, where it quickly succeeds such lichens and mosses as first established themselves, where even the small Nailwort refuses to vegetate.

The roots are brown, and occasionally clothed with a thick pile; the rhizoma is brown also, having a densely-covered skin or cuticle, which dries and peels off after a year's growth, leaving the rhizoma delicately smooth—a peculiarity rarely observable in ferns.

Leaves of the common Polypody are generally uniform; variations, however, occasionally occur, and should be noticed by the botanist: for this purpose we recommend a small book, with white paper, and pencil, to be carried in the pocket. As a specimen of the practice, we copy a note by Newman, transferred into his admirable "History of British Ferns:"—

" The common Polypody is somewhat parasitic, preferring the stem of a tree, or the half-decayed stumps of hazel and whitethorn bushes—over these its creeping rhizoma delights to wander. In the South of England it ascends the loftiest trees; and in Epping Forest I have often seen it ornamenting, with its bright green fronds, heads of the pollard hornbeams, when the wintry blast has stripped them of their summer verdure.

" In England this fern has insinuated itself into the mortar of our walls, houses, churches, and bridges; into our hedgerows also, and has become, in a manner, domesticated, yet does not enjoy such a perfect freedom as amid the humid, rocky, and shady dingles of Kerry and Wicklow."

Memorandums of the kind are readily made—they preserve the memory of favourite haunts, where grew the finest specimens, and when read to others may become suggestive of similar pursuits and pleasures.

This fern, though universally distributed, is, in our minds, particularly associated with the remembrance of Windsor Castle. Large tufts attracted our attention on the walls of the old Keep, where James of Scotland, the poet king, passed his dolorous captivity during the reign of Henry IV. They sprung, if we mistake not, from a fissure near the window from whence the captive looked down on the lady of his love,* when all unconsciously she gathered flowers in the small garden that extended at the base of the stern old fortress.

Hark to the rushing sound of the waterfall! its white foam may be seen among the trees, and far beneath our pathway. Tread carefully,—the bank is very steep, and though covered with brushwood and brambles, its sides are nearly perpendicular, and a false step might send you into the racing stream. Now we are safe, and can stand securely on the old bridge, which, as antiquaries tell, led from one kingdom of the Saxon heptarchy to another. Look over the parapet—the whirl of the eddying torrent is almost bewildering; but calmly grows that beautiful tuft of fern above the raging waters—the fern of waterfalls! to which the unmeaning name of Beech-fern (*Polypodium phegopteris* of authors) is applied. Why, we cannot tell: for this remarkably graceful and well-marked fern has rarely, if ever, been found beneath the shade of beech trees. It grows in damp localities, on dripping rocks, or in cavernous recesses, and within the spray of falling waters, where its wiry rhizoma, tough and uniformly creeping, often forms a network over perpendicular rocks.

The species are widely diffused. In this country, the mountainous districts of Lancashire and Yorkshire, Westmoreland, Cumberland, Durham, and Northumberland, are its favourite resorts; growing also near the town of Ludlow, renowned in Border history. In Scotland, Wales, and

* Daughter of the Duke of Somerset.

Ireland, not a mountain rill nor waterfall but owns this
favourite fern—on mountains, too, where clouds congregate,
and among huge unsheltered masses of rocks raged over by

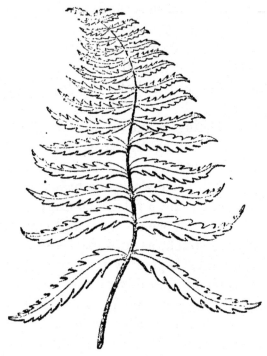

BEECH FERN.

storms. Those who visit the Pass of Glencoe and Loch
Katrine—the one with its dread records of crime and
misery, the other smiling in perfect beauty—may notice
this graceful fern as not unfrequent in both localities.

The Beech-fern, on account of its humid haunts, is
somewhat difficult to cultivate in a Fernery. Newman,
however, adopted a most ingenious expedient with com-
plete success. He suspended above the fern a vessel containing water, which he allowed to drop slowly on a stone
or flat tile contiguous to the plant; the fronds were in consequence kept moist by the mimic spray that rebounded
from the surface.

That man may justly be considered a benefactor to his species who opens or facilitates new sources of enjoyment, equally with him who causes wheat to grow where it never grew before. We hail the simple expedient of the dripping vessel as eminently calculated to induce the possessors of small outlets to beautify them with such ferns as require shade and moisture—to go forth among the lanes and woods at intervals of leisure, and derive enjoyment from the healthy recreations that are within their reach.

Think not to meet with the *Woodsia ilvensis*, and *W. hyperborea*, the *Polypodium arvonicum* of Withering, either in England or Ireland. In Wales the genus is rare, even on Snowdon; Dr. Richardson gathered it from a moist black rock nearly at the top of Clogwyn y Garnedd, facing north-west, and directly above the lower lake. Glyder Vawr (or the Hill of Tempests), and Clogwyn y Garnedd, in Carnarvonshire, afford isolated specimens. In Scotland it seems restricted to Perthshire, Ben Lawers, Forfarshire, and the Clova mountains.

For the sake of travellers whose summer or autumn excursions may lead them to those parts, we shall mention that the roots are long, fibrous, and brown; the rhizoma tufted, brown, slightly scaly; the young fronds, or leaves, appear in May, and continue green till September or October. The shape of the frond is linear, or strap-shaped—lanceolate, or spear-shaped—and pinnate, which term has been already explained; the pinnæ are attached by their stems only—they are indented, but not pinnatifid.

In the absence of specimens, we avail ourselves of Newman's admirable delineations of this rare fern.

Fig. 1 represents two pinnæ detached and magnified; the upper shows the masses of thecæ in their natural position; the lower exhibits the veins, and the points of attachment of the thecæ at their extremities, the thecæ themselves being removed.

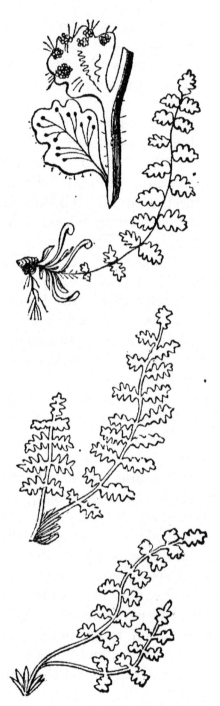

Fig. 1.

Sadler, who considers the *Woodsia ilvensis* and *W. hyper-borea* as distinct, thus characterizes them :—

*W. hyperborea.*—" Frond linear, lanceolate, pinnate—
under surface clothed with soft hairs; pinnæ nearly ovate,
obtuse at the base, unequally cuneate, nearly sessile (or
stalkless), obtusely lobato-pinnatifid; masses of thecæ
becoming nearly confluent, or running one into the other;
stripes smooth; rachis pilose."

*W. ilvensis.*—" Frend oblong, pinnate—hairy beneath;
pinnæ opposite, lanceolate, pinnatifid—the lobes oblong,
obtuse, lower ones spreading; masses of thecæ confluent;
stripes and rachis scaly-villose, or soft-haired. A small
portion of the rachis is naked, the veins irregularly distributed, the mid-vein is not to be traced without difficulty,
no single vein appearing superior to the rest, none reaching the margin, and each at its extremity bearing a mass of
thecæ."

We have recommended the introduction of different kinds
of moss in Ferneries, both on account of their beauty and
utility. Nature places them together; wherever the fern
spreads forth her ample fronds, there the simple moss
nestles beneath their shade—the one shelters her humble
friend from the fierce beams of a noonday sun—the other
gratefully protects the roots of her benefactress from
drought, by imbibing and retaining whatever moisture is
afforded by night-dews; a fact thus beautifully exemplified
in the following admirable lines :—

FERNS AND MOSSES; OR, THE LINKS BY WHICH SOCIETY IS HELD
TOGETHER.

There was fern on the mountain and moss on the moor—
The ferns were the rich and the mosses the poor;
And the glad breeze blew gaily—from heaven it came—
And the fragrance it shed over each was the same;
And the warm sun shone brightly, and gilded the fern,
And smiled on the lowly-born moss in its turn;
And the cool dews of night on the mountain-fern fell,
And they glisten'd upon the green mosses as well.

And the fern loved the mountain, the moss loved the moor,
For the ferns were the rich, and the mosses the poor.
But the keen blast blew bleakly, the sun waxed high
Oh! the ferns they were broken, and withered, and dry,
And the moss on the moorland grew faded and pale;
And the fern and the moss shrank alike from the gale.
So the fern on the mountain, the moss on the moor,
Were wither'd and black where they flourish'd before.

Then the fern and the moss they grew wiser in grief,
And each turned to the other for rest and relief;
And they plann'd that wherever the fern-roots should grow,
There surely the moss must lie sparkling below.
And the keen blast blew bleakly, the sun waxed fierce
But no winds and no sun to their cool roots could pierce.

For the fern threw her shadow the green moss upon,
Where the dew ever sparkled undried by the sun;
When the graceful fern trembled before the keen blast,
The moss guarded her roots till the storm-wind had pass'd.
So no longer the wind parch'd the roots of the one,
And the other was safe from the rays of the sun.

And thus, and for ever, where'er the ferns grow,
There surely the mosses lie sparkling below;
And thus they both flourish where nought grew before,
And both deck the woodland, the mountain, and moor.

---

## JUNE.

"THE cave was very chill, and damp withal,
　　And yet from out its lone depths shone a light
　　So pure, unearthly, radiant, that no eye
　　Might gaze unmoved upon it."

*Extracts from our Note-Book.*

*June 4th.*—Visited a lonely granitic cavern on Dartmoor.
The entrance was difficult, somewhat dangerous, in consequence of heavy rains, which had occasioned a considerable

fall from the roof. Unlike caves in general, which are often shaded with high trees, and having clear streams flowing near, of which the gentle murmur is blended with the song of birds and whisperings of winds among the branches, Argol's Cave looked damp and cheerless, and was associated in our minds with Druidic superstitions and fallen cromlechs. The naturalist who ventured nnadvisedly into that same cavern might have started back with some degree of apprehension; for out of its recesses gleamed forth a softened and beautiful light, enhanced by the twilight gloom that brooded within. This phenomenon, conneeted with peculiarity of structure, has its counterpart

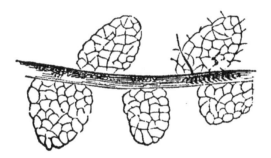

LUMINOUS INMATE OF ARGOL'S CAVE.—SMALL PCRTION,
HIGHLY MAGNIFIED.

among birds and insects, in the exquisite plumage of the humming-bird, and the burnished wings of the opal beetle; among stones, in the Labrador feldspar, or the precious opal of Hungary. The light thus wondrously gleaming from off the humid soil resembled a carpet of burnished gold, and was seen to the greatest advantage at the distance of a few yards, more especially when beheld from a favourable angle of vision. On near inspection, a variety of closely-scattered stones formed the basement of Argol's cave; they were covered with filmy irregular network, scarcely perceptible, from extreme delicacy of texture; and the light which served to betray the lonely, lovely' inmate of the cavern, was caused by rays of light concentrated by, and reflected from, the innumerable and inconceivably minute lenses of the leaves.

Botanists who visit Derbyshire during their summer excursions, may find the *Hypnum lucens*, or Shining Feather-moss, in the shady recesses of Rowter Rocks, a mile or two north of Winster. It grows in different situations, among woods, wet ditches, and on moist banks and rocks.

Withering speaks of it as having trailing branches, egg-shaped, pointed, and flat; leaves shining as if wet with dew; fruit-stalks an inch and a half long; capsules small for the size of the plant, somewhat oval, more or less nutant, dark brown; lid, spit-pointed; veil straight, sharp, whitish.

It occurs to us that the mild golden green light in Argol's cave is emitted by some species of conferva—unless, indeed, the shining Feather-moss is greatly deteriorated in size by its gloomy habitat.

Luminous plants produce an inexpressibly pleasing effect in their lone and desolate growing-places. Counsellor Erhman spoke of them with enthusiasm, when, having descended into one of the Swedish mines, he saw those vegetable glow-worms gleaming along its walls, or sparkling in some obscure recess. Caverns in the granitic rocks of Bohemia are often beautifully decked with a species of luminous moss; and our own coal-mines occasionally exhibit a light sufficiently clear to admit of reading by its aid. But nowhere, perhaps, is the effect produced by vegetable phosphorescence so exquisitely beautiful as in the mines of Hesse, in the north of Germany, where the walls of the air-galleries appear as if illuminated with a pale light, resembling that of moonbeams when stealing through crevices into some gloomy recess, from which all of vegetable beauty is excluded. None, in looking on the fairy gleams that pervade the Hessian mines, could imagine for a moment from whence the glancing lights proceeded; they would attribute them rather to some peculiarity in the strata of which the mine is composed, or to a kind of glistening spar that reflects the light of day. And yet that phosphorescence has a vegetable origin, an emblem, we have thought, of those gentle and retiring ones who render cheerful homes that have nought else of gladness to commend them, who shed the light of their pure examples over the moral gloom by which they are enveloped.

Dartmoor abounds in mosses of all shapes, and tints of green; those which delight in arid places find their dwellings on rocks and scarry banks, on cromlechs, and huge rocking-stones, fixed by time or accident; such as thrive best in moisture grow profusely beside the racing streams that water its wide expanse; old trees uphold to light and air others of a pendulous character; while not a few remain contentedly on maternal earth. Among these the *Hypnum sericum*, or Soft-ribbed Feather-moss—though occasionally affecting the trunks of trees and walls, carpets wide spaces on the moor, and exhibits the richness and softness of silk. When growing on the trunks and branches of trees, it may be scarcely removed entire, its small roots adhering so firmly to the bark; the leaves are soft and shining, slender, closely tiled, and ending in long hairs; the branches mostly point one way, with long, creeping, crowded shoots; the capsules are long, nearly cylindrical, but thickest at the base; and though minute, the fringe which surrounds the distinctly-formed mouth is white, with a beaked lid, and pale-coloured veil.

None among British mosses are more pleasing to the eye, both in form and colour, than the *Hypnum purum*, or otless Silky Feather-moss, common among woods, on heaths, and in meadows. Dartmoor is one of its common habitats; it grows equally in places open to the sun, and ade of those few memorial trees that linger in their loneliness and sterility, where once extended a vast brotherhood of stately trunks and intermingling branches.

The species may be readily known by its peculiarly sleek appearance, by its freedom from dirt, and its long cylindrical-winged scaly shoots, as also by being a span long in wet, but shorter in dry places. The leaves are thin and soft, smooth, and rather shining, and when dry, crumpled. This fine moss derives its name of *spotless* from the peculiarity already noticed.

*June 9th.*—A deeply interesting day. Gathered tufts of the Pendulous Feather-moss (*H. curtipendulum*) from a dwarf oak in Wistman's Wood, Dartmoor.

Wistman's Wood is associated with the most ancient records of our country. Its dwarf oak-trees, widely and wildly scattered, arise from out the interstices of granite masses that lie scattered in all directions, or else grow among them. Those trees, once stately and umbrageous, sprung most probably from the roots of such as were destroyed by fire, when many a widely-extended forest was cut down or burnt in winter, in order to dispossess the wild beasts and outlaws that sought their covert. Those stunted-looking trees, exposed to the continual action of bleak winds that rush howling past the precipitous descent on which they grow, have lost their upper branches, and look as if shortened of half their height; few, if any, are more than ten or twelve feet high; but though deprived of their natural beauty with respect to height, such branches as still remain have spread far and wide, twisting in the most fantastic manner, and festooned with ivy and creeping plants. Their trunks are also thickly imbedded in a covering of moss, and seem of enormous thickness in proportion to their height; but such is not the case, their apparent size is owing merely to the rich garniture that envelopes them. The moss by which they are invested, and which occasions stunted branches not larger than the wrist to equal in apparent size the trunks of giant oaks, is simply the Pendulous Feather-moss in its fullest development, growing from eight to twelve inches long, and producing thecæ in the greatest profusion. The same species also affects the trunks of beeches in woods ; it may be found on stumps in Enfield Forest near Southgate, and in Yorkshire ; on large stones scattered over the Marlborough Downs in Wiltshire, and on the heights of Snowdon.

Those who go in search of the Pendulous Feather-moss among the blasted oaks of Dartmoor will do well to remember, that though wearied with the toil of climbing the rocky path that leads to Wistman's Wood, they may not sit down to rest on the immense masses of granite around and beneath the trees, cushioned though they be with the thickest and softest moss, lest they should disturb a nest of adders. Of this an old man warned us who served as a guide across the moor. When thinking of the awful rites that were carried on by Druids among the groves of Dartmoor, when Baal and Ashtaroth were propitiated at early dawn, we could not help imagining that somewhat of the curse denounced on such unhallowed places rested on the site of the old oaks. " Serpents hiss there—the shepherd does not make his fold there, and the bittern screams amid its desolation." This was literally true. Nothing could exceed its sterile aspect, not a moving object met the eye—no sound was heard except the rushing of waters and the cry of a solitary bittern flying towards the valley of the Dart.

*June* 10*th*.—Sought for some time the *Hypnum bryoides*, or Bryum-like Feather-moss, which grows mostly in shady places, woods, and ditch-banks. Found at length a small brotherhood on the margin of a stream, which having forsaken its usual channel in consequence of a pebbly accumulation that hindered its onward progress, wasted itself upon the grass.

This kind of hypnum is very small, but distinguished by its capsules, edged at the mouth with a deep red fringe. Linnæus speaks of it as the smallest of the genus. The

shoots are two or three lines in length; the leaflets seven or
eight pair; fruit-stalks long or longer than the shoots, gene ·
rally solitary, reddish; leaves green, not pellucid; capsules
small, upright, oblong, green; veil very small, greenish;
lid scarlet; mid-rib of the leaflets pellucid.

The growing-place of this minute hypnum had much of
grandeur and sublimity. A vast plain extended on all sides,
looking in the distance like a desolate wilderness, or rather
as an ocean after a storm, heaving in large swells, and yet
presenting on a nearer view an almost endless continuation
of narrow valleys, of lofty hills and craggy rocks, strewn
either in their depth or far up their sides with enormous
masses of huge stones.

*Hypnum, triguetrum* and *undulatum*, equally affect the
dissimilar localities of Dartmoor. The first, Great Triangular
Feather-moss, abounds upon the roots of trees, and on dry
pebbly banks; the other, which bears the name of Waved
Feather-moss, prefers shady places, woods, and moist rocks,
or the top of Snowdon. This fine species differs essentially
from the rest of its genus, by having white, membranous,
and undulated leaves, and still more remarkably from all its
British congeners, by furrowed capsules. It is a span long,
lying flat; the leaflets are closely tiled, in a double or triple
series; the fruit-stalks long, slender, reddish; veil, straw-
coloured, with a brown spot at the end; the leaves are
tender, pellucid, smooth, shining, pale green, and not chang-
ing colour when dry.

The *Triguetrum* presents a widely different appearance,
indicating that it dwells low upon the ground. The branches
are unequal; the leaves broad, triangular, not keeled, tender,
pellucid, pointed, pale green when growing; involucrum
ridged, oblong, composed of reflected scales, sometimes two
or three together; fruit-stalks seldom more than an inch
high; capsules upright, thin, when ripe thicker, leaning,
and crooked. The whole plant spreads to a foot in length,
reddish, elastic, rising upwards, often growing upright,

although the branches frequently bend towards the ground, where their extremities strike and take root.

In this respect the humble Feather-moss resembles its **giant relative** (the *Ficus Indica* of Hindostan and Cochin-China), of which the lateral branches, sending down shoots which take root in the earth, compose a grove that often covers a wide area. We use the term relative—though one grows low, and may be trodden upon by every passer-by, and the other rises to a commanding height—because all vegetables are related, individuals of one great family; and what the Banian is to those who walk beneath its branches, whether Hindoo, Chinese, or European, so is the unassuming Feather-moss to insects that find a home and storehouse within its precincts. And if it be allowable to apply lines descriptive of that vast and peculiar Banian to its lowly brother, we may say with equal truth—

> " Many a long depending shoot,
> Seeking to strike its root,
> Straight, like a plummet, grows toward the ground;
> Some on the lower boughs which cross their way,
> Fixing their bearded fibres round and round,
> With many a ring and wild contortion wound;
> Some to the passing wind, at times with sway
> Of gentle motion swung."

*June 11th.*—A rainy day, yet passed pleasantly in spreading out the mosses which we had gathered between sheets of blotting-paper, and then laying upon them a heavy weight. Travellers are not provided in this respect, but we found an excellent substitute in a board borrowed from the landlady of our little inn, and this we covered with stones. Besides our note-book, we had taken the precaution of bringing another with blotting-paper sheets about the same size; and these, with tin cases for holding ferns or mosses, were all that we required for our botanical excursion.

*June 12th.*—Went forth again upon the moor. The

heavy rain of yesterday swelled many a wayside stream that flashed and sparkled in the sunbeams, and every blade of glass was surcharged with rain-drops.   Light-wreathing mists arose from off the moor—now flying up the hills—now chasing each other across the valleys—now seeming to open, and to present long vistas of rock and dingle, to which the beams of a cloudless sun imparted the magic of aërial tints and hues.

Our path led across one of the wildest portions of Dartmoor.   A tiny rill, rising in solitude and silence, had swelled into a stream, and the stream, as it flowed on, received the contributions of other streams, which at length in their congregated might became powerful and rapid. Athwart this stream stretched one of those primitive bridges which our rude forefathers had most probably erected ; the racing torrent rushed impetuously beneath, bounding over vast masses of stone, and falling in foaming sheets of dazzling whiteness.   But though old and lichen-dotted, and grey with age, that same old bridge was beloved of flowers ; the blue forget-me-not, the harebell, the golden saxifrage, with red crane-bills, and yellow snapdragons, looked down in . beautiful companionship on the hurrying waters.   Among these were several small ferns of considerable beauty—the Brittle-fern especially, which is fond of bridges, and establishes itself in the interstices of their stones.   This plant, equally elegant and fragile, is much sought after ; the genus to which it belongs was established by Bernhardi, who, having rejected the previous names of *Cystea, Polypodium, Aspidium, Cyclopteris,* gave it that of *Cystopteris fragilis,* or Brittle-fern.

On referring to Newman's " History of British Ferns," we observed that the name thus given appears to have been adopted by nearly all subsequent botanists—that the genus contains a limited number of species, which, although of wide dispersion, are restricted mostly to old walls and buildings, or dry stony places ; it may be that a torrent dashes beneath, or else beside them, as in the instance just cited ; but then the roots are firmly fixed among stones or pebbles. Such plants as pertain to this fragile genus are of small size, of erect and elegant growth, and remarkably brittle. " One species only belongs to this country, and on this much labour and ingenuity have been expended, in **the**

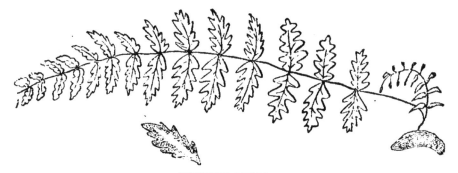

BRITTLE FERN.

hope that some of the most remarkable of its Protean fronds may be exalted to the dignity of species." Thus far the naturalist of Godalming, who has beautifully delineated many a wild occupant of wood or rock.

We rejoice to find the *Polytrichum aculeatum*, or Common Prickly-fern, in one isolated spot ; the more especially because this fern, though universally distributed, seems to delight in the neghbourhood of man, its favourite habitat being hedgerows, and the vicinity of cultivated fields ; when sown by winds on moors, or the sides of mountains, it rarely attains its full development ; and he who sees the Common Prickly-fern growing luxuriantly in a hedgerow, beside some way-side cottage, would scarcely recognise it in a desolate and unpeopled district. The same fern, however, grew luxuriantly in one of the wildest parts of Dartmoor. "Perhaps," we said, while looking at it, "men may have dwelt here in ages long past, and these friendly ferns, watching on the sight of some deserted home, continue as memorials of the past."

F

Few of the fern tribe are more pleasing to the eye.    The
young fronds become developed in April and May ; the apex
is circinnate (or bent backwards), and remarkably graceful; the pinnas are also bent backwards.    The whole plant
attains its full expansion in July, and the seed appears to
ripen in September.    Unlike the generality of their brother-

1. COMMON PRICKLY-FERN.

hood, Prickly-ferns are decidedly evergreen, and continue
throughont the year, uninjured by hard frosts and driving
storms ; they even linger on till late in the succeeding
summer, like the members of some patriarchal family
who dwell peaceably together.    Hence it happens, not unfrequently, that four generations are attached to the same
root, equally green and flourishing, yet naturally producing
leaves of a widely different character.    Even experienced
botanists have therefore been somewhat puzzled by dis-
similarities, which have suggested the idea of different
species.

    Four forms are assigned to the Common Prickly-fern.
Newman, however, considers that three only can be
reckoned ; and these he comprises in a single species, to
which he assigns the Linnæan name of *Aculeatum*, terming
the different forms merely varieties.    Thus :—

    *Var.* 1.—" Angular type : frond doubly pinnate ; pin-
nulæ ovate, bluntish, stalked, and auricled at the base; the
whole plant light, feathery, graceful, and extremely
flexible."    Figured as No. 1.

2. ASPIDIUM ANGULARE *of* 3. POLYPODIUM ACULEATUM
 *Smith and Hooker.* *of the " English Flora."*

*Var.* 2.—" Lobate type : frond doubly pinnate ; pinnulæ
pointed, decurrent, serrated--the formost of the lower pair

on each pinna very large, and pointing to the hard apex of the frond ; the whole plant rigid, heavy, compact, and unbending ; growing in general horizontally." Figured by No. 2.

*Var.* 3.—"Lonchitiform type : frond simply pinnate; pinnæ stalked, undivided, prickly ; habit weak, flexible, pendulous." Figured as No. 3.

*June* 11*th.*—Specimens carefully placed within the leaves of our blotting-book, and the book itself closely tied together. Farewell to Dartmoor! with its rocks and cairns, its rushing streams, and contrasted scenery.

" Dartmoor! thou wast to us in childhood's hour
   A wild and wondrous region. Day by day
   Arose upon our youthful eye thy belt
   Of hills—mysterious, shadowy, clasping all
   The green and cheerful landscape, sweetly spread
   Around our haunts; and with a stern delight
   We gazed on thee. How often on the speech
   Of thy half savage peasant have we hung,
   To hear of rock-crown'd heights, on which the cloud
   For ever rests, and wilds stupendous swept
   By mightiest storms; of glen, and gorge, and cliff
   Terrific, beetling o'er the stone-strew'd vale :
   And giant masses, by the midnight flash
   Struck from the mountain's lofty brow, and hurl'd
   Into the foaming torrents !"
                                        CARRINGTON.

## JULY.

"Green the land is where my daily
Steps in jocund childhood play'd—
Dimpled close with hill and valley,
Dappled every close with shade;
Summer-snow of apple blossoms, running up
From glade to glade."

ELIZABETH BARRETT.

BACK to our own cottage home, mid glens and waterfalls, where grow most of the ferns and mosses which we found on Dartmoor! Who that have travelled forth in quest of knowledge,—Nature's pilgrims, visiting each shrine and dwelling-place of the rarest or loveliest of her offspring,— do not remember the delight with which their books containing specimens were opened, and how vividly arose before their mental view the rock, or stream, or bank, the solitary glen or woodside, from whence they gathered the plant, the fern, or moss, which they had journeyed far to find?

Such were our feelings as we carefully removed our ferns and mosses, looking on them with somewhat of pride, when noticing that not even the smallest pinnæ had been injured in drying, nor yet a single root distorted from its place. Then came the pleasure of arranging them, of assigning the mosses to our moss-books, and the ferns to occupy the pages of a larger volume; the writing of their names and where they grew, with such botanic memoranda concerning cromlechs and old rocking-stones, ancient bridges across racing torrents, and the trees of Wistman's Wood, as brought up pleasant memories of our rambles on that wild moor, which has no parallel in British scenery.

We gave you, botanical friends, drawings of the ferns, but one wet day did not suffice for delineating such mosses as we gathered.  Accept them now, for we are at home again, and have abundant leisure for the employment of our pencil.  But first let us consider the component parts of these small plants, and reflect for a brief space on their admirable construction.

Observe those urn or vase-shaped vessels, on the summit of stems or peduncles arising from among green foliage! They contain innumerable seeds, which are either sown by

winds when ripe, or scattered immediately upon the earth, and the base of the upholding stem, closely examined, discovers a sheath of scaly leaves.  Those leaves were all-important to the welfare of the plant; before the peduncle springs up to light and air, they serve as a kind of calynx to protect the embryo fruit, and bear the name of sheath or perichetium: *a* capsule, *b* pedicle, *c* sheath.  Observe, also, two important parts connected with the capsule, before the period when it splits open and its contents are scattered abroad—*d* the operculum or lid, which closes the mouth, and *e* the calyptra or veil, which covers both the lid and capsule like a conical roof; *f* is the fringe or peristome (*peristomium*), which becomes apparent when the seeds being fully ripe no longer require protection, and the casting off of the lid discovers the opening, which is generally ornamented by a circle of saw -like teeth.

Mosses are exquisitely varied, and many of the stems which uphold flowers containing one or .more stamens, but no pistils, have a star-like appearance at the top. Beautiful in their minuteness, they form interesting objects for the microscope; their serrated and ribbed leaves are uniformly thin, pellucid, and veined like network; the roots are fibrous; and not unfrequently both stems and branches throw out fine roots whenever they come in contact with the earth, or any supporting substance. Some few are so exquisitely fine as to require a high magnifier in order to discover their various parts; yet even these have stems and capsules, and occasionally present the appearance of fairy trees; more generally they are from one to three inches high, though the great Hair-moss (*Polytrichum commune*), and some kinds of bog-moss, which grow in watery places, are nearly two feet in height.

Obtain a microscope, if you wish to study mosses attentively, and to become acquainted with the most delicate species; one with a simple lens will magnify sufficiently, and for this purpose we recommend Ellis's aquatic microscope, according to the suggestion of Dr. Drummond. If, on the contrary, you are contented with less close investigations, a common magnifying glass, purchased for a few shillings, will suffice. Thus will a new world be opened to your minds; and many a moss or lichen-dotted stone that would otherwise be passed unheeded will become suggestive of much that is worthy of regard.

Ireland has been deservedly called the paradise of mosses; some of the most beautiful species grow there abundantly beneath the shade of trees in marshes, and beside waterfalls; the *Fontinalis antipyretica*, or Greater Water-moss, especially, which delights in the neigbourhood of cataracts, and flourishes the most where a racing stream eddies eside its growing place, and tosses on high its billowy spray.

Such as grow in tropical regions prefer the shade of rocks, especially when assigned to alpine heights, where the temperature of the climate is rendered moderate by their elevation.

The brotherhood, like most of their cryptogamic relatives, are tenacious of life; even when apparently dead, a shower presently revives them. They thrive, too, in places unfavourable for general vegetation; and hence, wherever a wandering sunbeam enters, or breezes gain access, some tiny moss, it may be, finds a home. You may discover them equally in cold, damp caverns, and in fissures among rocks, or on walls open to the sun.

Wonderful it is, that when some species are subjected to great drought at the time appointed for ripening their seeds, they acquire the property of absorbing and retaining moisture, like succulent plants; the process of ripening consequently advances rapidly, even if the heavens deny rain, and the earth is hard as iron. But though assigned to different regions of the globe, and various growing places in this country, they generally affect temperate and cold regions, and often, in companionship with lichens, present the last trace of vegetation towards the limits of perpetual snow. Their uses are multifarious—they protect young plants and seeds during the heat of summer, and in the depths of winter, and form retreats for insects and small animals. Travellers who explore the vast forests that extend far north, relate that the trunks and branches of the trees are covered with mosses, especially on the northern side; and that by means of these natural indications, those who traverse them in quest of animals readily find their way.

Abundant in mountainous countries—rare on plains; somewhat restricted, also, in its localities, the only station where this moss has been discovered in the eastern angle of Great Britain, is on the sandy waste near Yarmouth. The dark and almost blackish green, cylindrical, and straggling, though somewhat pinnate stems of the shining Feather-moss, are very peculiar. Hooker speaks of gathering it in the wildest parts of Dartmoor, at least eight or ten inches long, and in a fine state of fructification.

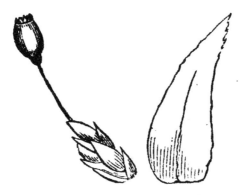

HYPNUM CURTIPENDULUM.

*H. Purum.*—Leaves closely imbricated, oval, with a very short point, concave, their nerve reaching half-way up; capsule ovate; lid conical.

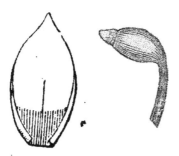

H. PURUM.

*H. Lucens.*—Leaves ovate, nerve disappearing below the point; fruit stalks long; capsule ovate.

We cannot, perhaps, do better than occupy the remaining pages with observations transcribed from our note-books, relative to such ferns as we have gathered in our botanic rambles, whether near our home, or in distant localities.

H. LUCENS.

The Marsh-fern or *Lastræa thelypteris; Polypodium thelypteris* of Withering, Lightfoot, Berkenhout, and Hudson, and bearing at least four names assigned by botanists, is one of our rarest and most local species. It abounds in wet and marshy grounds, and moist woods and bogs, where its black, slender, and wiry rhizoma often creeps to a wide extent. The roots are also dark, fibrous, and, in some instances, very long ; they frequently penetrate to a considerable depth, while the rhizoma spreads widely and horizontally.

This local species is unknown in Wales and Scotland, and in Ireland is generally believed to be restricted to the county of Antrim, near the north-east coast of Lough Neagh.

> " On Lough Neagh's banks, where the fisherman strays,
>    When the clear cold eve's declining,
> And sees the round tower of other days
>    In the stream beneath him shining."

Throughout England, the localities of the Marsh-fern are widely spread, but universally of a moist and humid character. Learmouth Bogs, in Northumberland, are consequently its favourite resort ; as also marshy places in Cheshire ; Whittlesea Mere, Cambridgeshire ; Filsby, Ormsby, Bolton Bay, Horning Marshes, Kent ; Norfolk ;

Ham Pond, near Sandwich, where it luxuriates in the meadows, and banishes all other ferns from the neighbouring wood ; those who visit the Isle of Wight, will find this marsh-loving fern near Fresh-water Gate.

For the sake of botanists who may be somewhat puzzled

MARSH FERN.

when referring to Withering's *Arrangement of British Plants*, pp. 995, 996, we subjoin a concise description of the Marsh-fern, extracted from the *History of British Ferns*, being anxious rather to avail ourselves of the aid afforded by that valuable work, than to rely exclusively on our own judgment.

Some of the fronds are barren, others fertile. The former rising from the bog in May, the latter in July—they both disappear with the first frosts of winter. The frond is lanceolate and pinnate; the lowermost pinnæ are shorter than the third and fourth pairs; they are attached by their

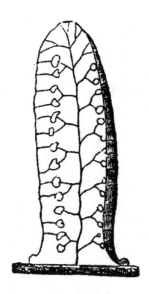

VEINS AND THECÆ.

stalk only; about one-third of the rachis is without pinnæ; the pinnæ are pinnatifid, the pinnulæ rounded, and always entire; the whole plant is erect, very slender, delicate, and fragile; it is of a pale green colour; in size varying considerably, in some instances even to four times the usual dimensions. The fertile fronds are uniformly larger and of stronger growth than the barren.

The lateral veins are alternate; they are forked almost immediately on leaving the mid-vein, and each proceeds to the margin of the pinnula, bearing a circular mass of thecæ almost directly after the fork. Each mass of seeds has, in an early state of the plant, a small subreniform indusium attached on one side to the vein, at the point to which the stalks of the thecæ are affixed.

Green **grows** the Mountain-fern, on many a lordly ridge
**and** herbless crag, in quarries, and where clouds congregate:

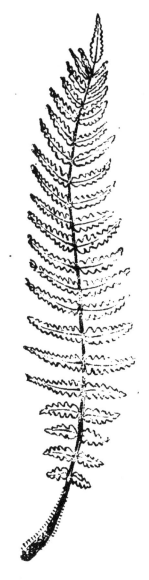

MOUNTAIN FERN.

**throughout** North Wales, therefore, high **mountain** ranges
are often covered with its widely-spreading fronds; and so

abundant are the brotherhood among the highlands of
Scotland, as often to take the place of the Eagle-fern. We
have gathered it in many of the northern English counties :
on the Clee Hills in Shropshire, and near Matlock. Northamptonshire, though generally unfavourable for the growth
of ferns, cherishes this wild species in some of its localities,
and botanists speak of having found it both in Hereford-
shire and Oxfordshire. In Wiltshire, on the contrary,
Mountain-ferns are scarce, though indigenous to Somersetshire, Kent and Sussex, Norfolk, Essex, and Middlesex,
among which Tunbridge Wells, Epping Forest, and Hamp-
stead Heath, are very accessible to Fern collectors.

Withering speaks of the Mountain-fern under the appel-
tion of *Polypodium Oreopteris;* Hudson and Berkenhout
give it the simple name of *Fragrans;* Bolton, that of
*Thelypteris;* other botanists, *Aspidium.*

A peculiar characteristic pertains to this interesting
species, namely, yellowish resinous glands, spreading over
the back of the leaves ; they inhale a fragrant scent, which
induced Hudson to regard them as the *P. fragrans* of
Linnæus.

This graceful fern is pleasantly associated in our remembrance with the Clee Hills in Shropshire, covered with wild
thyme, and ranged over by innumerable bees, when gathering their honey harvests. We have counted twenty, or
even thirty leaves, encircling a common centre ; and among
them the wild hyacinth often lifted up her peerless blue-
bells, as if grateful for their shelter. It was pleasant to go
forth from day to day, and watch the gradual unrolling of
the leaves ; to observe how accurately the pinnæ were
placed at right angles with the rachis ; while the swallow,
obedient to the voice of spring, darted swiftly through the
heavens, and the skylark soared and warbled in his up-
ward flight. Ferns and blue-bells, the lark and swallow,
were then to us as monitors, noting the beauty and the
order of creation; each in its little sphere obedient to the

elements, and communing with them ; each one unfolding or arriving, or attuning its heaven-taught minstrelsy in accordance with unalterable laws, which have never changed since the round earth emerged from chaos in its glory and its beauty, and went on its way rejoicing amid kindred worlds.

Little or no variation occurs in the figure of the frond or leaf. It is uniformly elongate, lance-shaped, regularly pinnate, acute or sharp-pointed at the end, and gradually diminishing from about two-thirds of its length to the base, the lower pinnæ being remarkably short—a peculiarity which sufficiently distinguishes it from all other ferns. A small portion only of the rachis is bare, and covered with scales. The pinnæ are linear, acute at the apex, somewhat distant, deeply pinnatifid, and affixed to the rachis only bv their stalks; the pinnulæ are rounded, and slightly crenate.

THECÆ OF THE MOUNTAIN FERN.

The veins present a simple alternate series, ceasing before they reach the margin ; circular and naked thecæ are borne by each, and yet occasionally the veins divide nearly at their termination, in which case each division reveals a separate mass. These masses, varying on either side from five to ten in number, form regular and nearly marginal series, extremely ornamental, and presenting pleasing objects for the microscope.

––––––––––––

## AUGUST.

" I cannot but think the very complacency and satisfaction which a man takes in these works of Nature, to be a laudable, if not a virtuous habit of mind."

ADDISON.

THE geography of plants is a subject of the deepest interest, suggestive, too, of pleasant thoughts, and often bringing before the mental view remembrances of bygone days, when rambling through woods and vales, by streams and over breezy commons, some long sought-for plant was discovered in its own lone habitat. We speak not of that perfect order which pervades the universe, concerning the assignment of vegetable tribes or families to regions far remote—of the bread-fruit and the palm to sunny climes, and fir-trees to cold inhospitable lands—of plants invaluable to mariners, among otherwise sterile rocks, in seas where men go in quest of whales. Our attention is directed rather to exemplifications of the same arrangement, conspicuous in this country and its sister island; and of this the *Trichomanes speciosum* of Willdenow, or the *T. alatum* of Withering, affords a striking instance.

The Bristle-Fern, for such is its familiar name, is one of the most interesting and local of British species. Newman reports it as utterly unknown in England, Wales, and Scotland—as growing sparingly in the county of Wicklow, at Hermitage Glen and Power's Court Waterfall, though at neither of these localities has more than a single specimen been discovered; luxuriantly near Youghal, Glendine, in the county of Cork, and equally so at Turk's Waterfall, near

Turk Lake, Killarney. The same naturalist gathered specimens of great beauty to the left of the site whence tourists
obtain the first view of the fall. He tells us, that about fifteen yards higher up the stream, a rocky bank projects into
the river, which can only be approached by leaping from
stone to stone, amid the racing torrent and deafening roar

BRISTLE-FERN.

and spray of the descending waters. Friendly roots and
branches aided the adventurous botanist in climbing up to
the wild growing place of the beauteous Trichomanes, who
dwelt securely on a rocky bank, her dark green fronds
dripping and begemmed with sparkling drops, shining and
glittering in the sunbeams. Thus cherished, amid rocks
and waterfalls, grows the fern of which we speak, sought for
in vain amid green fields and on village commons, where
the eye-bright and the cistus, the twayblade and bee-flower,
open to the purest air of heaven.

The roots of this rare plant, equally with its rhizoma, bear
a considerable resemblance to those of the *Polypodium vulgare*, or common Polypody. The latter is black, velvety,

G

tough, and of great length, occasionally many yards, and often forming a kind of net-work on the perpendicular surface of damp rocks, which afford no kind of hold to the widely-diffused roots. This is the character of the fern when favourably situated for its full development; other and smaller plants are mentioned as possessing more root and less rhizoma, the former of which were fixed in a thin layer of earth, where a brotherhood of mosses grew luxuriantly.

The leaves appear in summer, yet rarely before August, and seldom attain to maturity till late in October, when the fronds of the previous year may be generally seen, dark coloured, but unfaded. Botanists who visit tropical regions speak of the genus *Trichomanes*, as comprising many rare and beautiful species. Go into the native woods, say they, of the West India islands, and observe how gracefully the trunks of noble trees are clothed with this elegant natural drapery. If you have hitherto passed unheedingly by the ferns of your own country, regardless of their symmetry and exquisite variety of form, it may be because of their unobtrusiveness; but you cannot withhold your admiration from those of exotic growth. They look down upon you from beetling crags—they cast deep shadows over your path— they often rise to a commanding height, waving and quivering far above your heads, while such as clothe the trunks of giant trees, present in their forms and leaves some of the most beauteous developments of vegetable life.

When speaking of the British species, Sir T. E. Smith describes the masses of thecæ as being roundish, terminal, and imbedded in the margin or segment of the frond. Indusium urn-shaped, in texture similar to the frond, and continuous with it, forming one leaf dilated upwards, and opening outwards, permanent. Thecæ several, sessile, crowded at the base of a permanent cylindrical common receptacle, whose capillary naked point projects beyond the cover, each roundish, with two valves, and bounded by a vertical-jointed ring.

Such is the definition of the generic characters as given in the *English Flora*. The author of a *History of British Ferns*, to whom we have frequently referred, furnishes, however, a different description of the Bristle-fern, which, in justice to our readers, we shall transcribe, remembering how singularly different species of this interesting family are affected by wet or dry summers, or by dissimilarity of growing places.

"The mass of thecæ is attached to the centre of a vein, after its ultimate division, and invariably to that one which is situated nearest the mid-vein of the frond, pinna, or pinnula, as the case may be. At the attachment of this mass of thecæ, the wing loses its green and semi-membranous appearance; its cuticles separate, and form an elongate, cup-shaped receptacle, which includes the mass of thecæ. The vein itself, after bearing the thecæ, runs through the receptacle, and projects considerably beyond its extremity, in the similitude of a bristle."

The Scaly-hart's-tongue (*Ceterach Officinarum*) has ever been my delight. It has no beauty to commend it, as

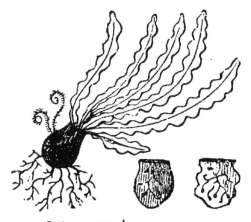

SCALY-HART'S-TONGUE.

figured above; but when growing on walls and rocks, in company with the harebell and small snapdragon, its frequent companions, there is something indescribably pleasing in the aspect of this unassuming fern. Perhaps we love it the more because of gathering it in our youth, on many a summer holiday, from an old wall at Totnels, near Pains-wick, Gloucestershire. We knew little of rocks and ruins then ; but a legend of deep interest was associated with that lichen-covered wall, and its contiguous mansion, and it threw around the whole a kind of romance which vividly affected our young imaginations.

Since then, we have gathered the same fern in widely different localities, but never without a feeling somewhat akin to melancholy : for who can look back unmoved on the haunts of childhood ; and what powerful awakings up of past realities are often elicited by the simplest fern or flower !

> Ye green ferns and flowers,
> Beloved in past hours,
> Ere the young heart had yielded its gladness ;
> We gaze on you still,
> By the gush of the rill,
> In the depth of our spirit's lone sadness.

Thus sung a mournful poet ; but let us rather rejoice in the beauties and wonders of creation ; and if, perchance, while looking at some plant or flower, such as the eager hand of childhood gathered in its gladness, and tendered as a tribute of its love to those whom now the earth owns not, their freshness and up-springing may well remind us of that glorious morning when parted ones shall meet again, and this " mortal shall put on immortality."

Unlike the Bristle-fern, which is locally restricted, the fern that gave rise to this digression very generally occurs in the south-western counties of England and Ireland, al-though of rare occurrence in the midland and northern counties ; and in Scotland is to be met with only at Dun-donald, near Paisley, and at the carse of Gowrie, according to the testimony of Dr. Young, and Mr. James Macnab, curator of the Horticultural Society's Experimental Garden at Edinburgh.

We shall transcribe the various localities of the Scaly-hart's-tongue, for the sake of young botanists who may either reside in their vicinities, or visit them during summer and autumnal rambles, prefacing our notices with the observation that in this country it has apparently become naturalized in the interstices of walls and ancient buildings, striking its small roots into the mortar, or accumulations of vegetable mould, scant though they be, yet sufficing for the requirements of such a tiny plant.

*England.*—" Yorkshire, very rare; a few fronds so labelled are in *Herbaria.*" On Ragland Castle, and Tintern Abbey, in Monmouthshire; diffused through various parts of Somersetshire, Devonshire and Cornwall; in the former, the neighbourhood of Bath, Bristol, Wells, and Langport, are its favourite growing places. In Berkshire, Pusey, near Faringdon; in Hampshire, the walls of the city of Winchester; in Kent, Tunbridge Wells, Maidstone church, Swancombe church, Shorn church. Old walls in Hereford and Leominster reveal the same interesting plant; those also of the Abbey church at Malvern; Ludlow Castle, in Shropshire, inseparably associated with the *Mask of Comus;* as likewise walls in the vicinity, are occasionally varied with small tufts. The dry fissures of rocks, at Dovedale, Cheddar, and those of a rock beside the road between Carnarvon and Bangor, are believed to be the only places where it occurs in its natural habits.

*In Wales.*—Walls and rocks near Bangor, and the neighbourhood of Swansea, are acknowledged localities; as caves in Holyhead mountain.

*In Scotland.*—Dundonald and the Carse of Gowrie, according to Hooker.

*In Ireland.*—Counties Dublin, Wicklow, Kilkenny, **Tipperary**, Cork, Kerry, Clare, and Galway.

Roots of the Scaly-hart's-tongue are endowed with **the**

singular property of penetrating mortar; thus admirably is
even the humblest plant adapted to the site which it is designed to occupy. The green leaves or fronds appear in
May and June, and attain perfection in August; they continue uninjured by frosts or rains, and are uniformly
fertile.

This small fern is easily distinguished, even when growing
among such as are deteriorated by ungenial soil, or ex-
posure on walls and rocks. The rhizoma is tufted, brown,
and scaly, and a small portion of the rachis alone is naked,
beset more or less thickly with pointed, chaffy scales.
"The frond is linear, elongate, and pinnate, or pinnatifid;
the pinnæ are attached to the rachis by their entire base,
and are sometimes also connected with each other; they are
obtuse, rounded, and crenate; the entire under-surface of
the frond is covered with brown, pointed scales, thought by
some botanists to be analogous to the indusium of other
ferns."

"The side veins are few in number, alternate and irregularly branched; they terminate before the margin of the
pinna, and are united at their extremities, dividing the
pinna into numerous compartments. The anterior branch
of each lateral vein bears an elongate mass of thecæ, fixed
apparently to the back of the vein, and seeming as if forced
aside by the surrounding scales." Occasionally they are
attached to a lateral vein, which in each pinna runs parallel
with the rachis.

Four names have been given to the Scaly-hart's-tongue
by different authors. Hooker, Mackay, and Francis call it
*Grammitis celerach ;* Linnæus, Withering, Hudson, Lightfoot, Bolton, Berkenhout, *Asplenium celerach ;* Smith and
Galpine, *Scolopendrium ceteraceh ;* Willdenow, *Celerach
officinale.* Such are the various appellations by which this
pleasing little fern is known to botanists; but however frequent on inland rocks, and old walls cemented by mortar
mixed with clay, we are informed that it is becoming very

scarce in places frequented by fishermen, being in great request for bait in rock-cod fishing.

Dioscorides celebrates the medicinal virtues of this fern,
as affording an almost universal panacea.

The *Asplenium trichomanes,* or common Spleenwort of
botanists, is very generally diffused. You may gather it

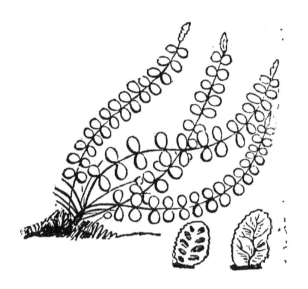

COMMON SPLEENWORT.

from moist rocks in mountain solitudes, on old walls, and
beside rushing torrents, on banks, n hedgerows, and from
the gothic windows of dismantled abbeys, where a scanty
supply of earth, in some small crevice, affords a resting-
place for its tiny roots. Newman once observed the same
plant in the valley of the Wye, near the small town of Bualt,
growing in such profusion on a bridge as to form a continuous covering of green; and truly, said he, "There is
scarcely anything in the vegetable world more beautiful
than such a sight." He recommends its cultivation, and
assures his readers that the effect produced on the bridge by
natural growth, may speedily be realized at home. Imagine
an unsightly wall, or a heap of stones still more unsightly,

in some dull corner unvisited by sunbeams, depressing the
mind of him who is daily constrained to look upon them;
contrast with this the pleasure of going forth into the woods,
where, perhaps, some time-worn ruin recalls the memory of
feudal greatness, and gathering from its walls tufts of the
common Spleenwort to plant among the interstices or stones.
If woods or memorial ruins are far away, you may seek for
the same fern in other localities; and to this succeeds the
pleasure of planting, watering, and watching the unfolding
of one small leaf, then another, till, as months pass on,
the bare wall or stones are mantled with a luxuriant vegetation.

The roots are black, tough, and penetrating; wherever
the smallest fibre insinuates itself, there the common Spleenwort makes good his footing, it may be in rock or wall,
exposed to the fury of fierce winds or scorching sunbeams,
or within the spray of waterfalls; this matters not.    The
Spleenwort is a citizen of the vegetable world, appearing in
May and June, arriving at maturity in August and September,
and remaining green throughout the winter.

Would you seek to place this fern among your specimens,
observe that the " rachis is naked for a third part of its
length, smooth, shining, and black throughout; that the
frond is narrow, linear, and simply pinnate; the pinnæ
dark green and very numerous; irregularly ovate, obtuse at
the apex, and more or less crenate at the margins; that
though they are usually distinct and distant,˙ occasionally
they are crowded, and each recumbent on the one preceding
it; that, moreover, they are attached to the rachis solely by
their stalks, falling off like the leaves of phanogamous plants
when the frond approaches decay, and leaving the rachis a
bare denuded bristle."    In size the common Spleenwort
varies considerably; at one time presenting a fairy-like
app·arance, at another one of considerable dimensions.

Observe, also, that the lateral veins are forked shortly
after leaving the mid-vein, the anterior branch bearing an

elongate, linear mass of thecæ, almost immediately after the fork; this mass is at first covered with an elongate, linear, white membranous indusium, which, as the thecæ swell, becomes obliterated; the black masses likewise become nearly confluent in two portions: they, however, rarely unite over the mid-rib, though ten or twelve in number.

This common fern has no pretension to medicinal virtues. Withering merely speaks of it as being generally substituted for the true Maidenhair in making capillaire—a syrup which, when perfumed with orange flowers, is considered an agreeable beverage.

Fern collectors must now begin to gather such specimens as they desire to possess; and where is there a fern that does not amply repay the trouble of preserving? Unlike flowers, which often become discoloured, or lose their vivid tints and graceful forms, these plants dry well, and retain the symmetry of their leaves and pinnæ; suggestive, too, are they of pleasures yet to come, among green woods and lanes, and bringing to remembrance many a country walk, when the cuckoo's song was heard, and ferns began to unfold in sunny brakes.

> " Let Fate do her worst, these are relics of joy,
> Bright beams of the past which she cannot destroy;
> Oh ! long be our hearts with such memories fill'd,
> Like the vase in which roses have once been distill'd ;
> You may break, you may ruin the vase, if you will,
> But the scent of the roses will hang round it still."

Such ferns as you wish to preserve should be gathered in dry weather, yet not when the pinnæ are slightly curled by the heat, which is frequently the case, but when the whole plant is fully developed. Lay the specimen thus gathered between several sheets of blotting-paper, taking care that even the finest root or pinnæ remains uninjured. Subject the whole to a considerable pressure, and let the fern continue unexamined for a few days; then carefully remove the weight and blotting-paper, lest any degree of humidity should require a change of paper. This, however, rarely occurs, except in the case of the Hart's-tongue, a water-loving fern, or that of the Moonwort. Replace the weight, and let your specimens remain in a dry place for a month or six weeks, when they may be safely removed to the pages of your fern-book, upon which it is desirable to tack them with a fine needle and silk or thread, in two or three places, according to the size of the fern. Do not forget to set down the places where grew your specimens, and any historic or biographic memoranda that can be comprised in a few lines. If, also, you meet with poetry befitting the ferns or their localities, such quotations will render your book still more interesting.

When the stems of the ferns are unusually thick, we have found it desirable to place on either side folded sheets of blotting-paper, so that the finer portions of the fronds may sustain an equal pressure.

---

## SEPTEMBER.

How beautiful are ripening fields of grain,
Varying the landscape. And how fair the scene
Of hill and dale, and woodland spreading wide;
With cottage homes, and village fanes, that lift
Their spires to heaven.

WHY is it that the Brake-fern grows profusely in some parts of Kent, whilst almost every other fern is wanting? that banks and hedgerows in the neighbourhood of Sydenham, especially, are profusely feathered with this interesting species? We have spoken elsewhere concerning the *Pteris aquilina*—of its wide diffusion, and association with memorial sites and ruins; but nowhere have we seen it more pleasingly localized than in a sloping field which extend from the turnpike road on Sydenham Hill towards the village. Ripening ears of grain grow luxuriantly on either side the pathway; to some they might have formed a rustling canopy, but to us they presented the semblance of tall stems, up-lifting their luxuriant heads to air and light, bringing forcibly to mind the vivid description of St. Pierre, who loved to lie down among the grass and corn, and observe the dappled insects that darted merrily in all directions.

Half-way down, the view was beautiful; full in front arose the stately tower of Upper Sydenham church, and in the distance the tall spire of Penge church was seen among the trees. Far as the eye could reach were hills and woods, and in the middle distance corn-fields and pasture land, with sheep and cattle. While lingering to admire the lovely landscape, comprising in its length and breadth much that is especially characteristic of English scenery, we observed on our left a space of broken ground covered with ferns and furze, and encircled with rustling grain; on the verge of this wild spot, and nestling among the corn, stood a small brotherhood of Brake-ferns, so comparatively small and delicately formed, that the eye, in looking on them, seemed to behold vegetable prototypes of those modest and retiring ones who shrink from the rough paths in which others of less gentle mood delight to venture. But our problem has not yet been solved. Who may tell the reason why, in a land thus favoured with hills, and dales, and sunny glades among the woods, the Brake-fern alone is found? This is one of Nature's mysteries; or it may be that we are in-structed by the wonderful arrangement of the vegetable world, that all things have their prescribed limits—that, moreover, the smallest plant or fern has a lesson inscribed on its leaves, which the passer-by will do well to read, bidding him take note, that each one is endowed with qualities which represent somewhat in the moral world. Thus, for instance, the geographic arrangement of trees, and shrubs, and flowers—of herbs and parasitic plants, admirably exemplifies the assignment of different races, among men, to various portions of the earth; while such as mostly beautify the trunks or branches of old trees, hint instructive thoughts of mutual benefits, and of that dependence on each other which renders every individual a benefactor to his kind.

The *Athyrium filix-femina*, or Lady-fern—the *Polypodium filix-femina* of Lightfoot, Bolton, and Withering—though growing profusely in moist and shady places, about rivulets, and on heaths, yet frequently adorns the aged heads of

LADY-FERN,

pollard trees, and often springs from out of the hollows wrought by time, or woodpeckers. This species, one of the most elegant among British ferns, though universally yet not equally distributed, is pleasingly associated in our remembrance with a wild and solitary place in Gloucestershire, which botanists may visit with advantage. That place is called Custom Scrubs; its locality is beside the old road from Stroud to Cheltenham, where the traveller, having ascended a considerable eminence, passes a fine beech wood, and looks down on the pretty little town of Painswick, beautifully situated on the declivity of a hill, the summit of which is crowned with an old Roman encampment thrown up by Ostorius. The road passes a series of valleys, renowned in history as the last strongholds of the ruthless Danes in the time of Alfred; and on the verge of the most remote and solitary, stands Custom Scrubs, with its rude cottages, and profusion of dark junipers. There grows the Lady-fern, a name expressive of its graceful and fragile form. Ray applied the term to our common Brakes, but Linnæus, with that delicate perception of whatever was most appropriate, assigned to it the one of which we speak.

Two distinct types of form pertain to the Lady-fern, and may be thus described :—

1. *Flattened type.*—" The fronds are broad, heavy, and drooping, and often of considerable size, perhaps even from three to five feet in length; the pinnulæ are perfectly flat, with all·their cuttings clearly displayed; and the masses of thecæ seldom, if ever, become perfectly confluent. Plants of this type vary infinitely in the cutting of the pinnulæ; also in the colour of the rachis, which is green, or inclining to red, purple, or even brown."

2. *Convex type.*—" The fronds are more narrow, rigid, erect, light, and feathery, of a smaller size, but still occasionally reaching from two feet to thirty inches in height; the pinnulæ are convex, the margins uniformly bent downwards; the masses of thecæ crowded and confluent; the rachis somewhat pellucid, and very brittle. This type is generally pale green, sometimes nearly white, occasionally of a pinkish tinge, and even nearly as red as coral."

Observe, also, that in these two very marked varieties, the one with broader segments of dark green hue, and having a rachis of pale purple, is less common than the variety of which the segments are of a more delicate texture, and the frond itself of a pale green. The latter varies considerably in size, according to soil and situation. In damp and shady places, beside streams, and on dripping rocks, it becomes the *Filix femina* of English botany, according to Professor Don, in the *Transactions of the Linnæan Society*, vol. xvii. p. 436 ; in more open and exposed situations, the *Aspidium irriguum* of Smith. But in ne'ther of these states is it to be regarded as a distinct form. Newman further mentions, that the margins of each pinnulæ are folded together, in that variety of which the segments are of a more delicate texture —and that they are so convolute as nearly to meet, which character causes each pinnulæ to look very narrow from above ; whereas, such as pertain to the other variety are spread out, and flat—the serrature, or lobes, being perfectly displayed.

The root is fibrous, black, and wiry ; the rhizoma is ver-tically elongate, rising, in some specimens, several inches above the surface of the ground, even occasionally to a foot in height, and thus evincing "considerable proximity to the Dicksoniæ, and other tree-ferns." The fronds appear in May ; and the bending downward of the apex, after the fashion of a shepherd's crook, causes them to resemble those of the Filix mas.

Examine the frond. "In form it is somewhat lanceolate and pinnate ; the pinnæ are linear, more or less crowded, acute at the apex, and regularly pinnate ; the pinnulæ are very distinct and distant, either deeply serrated, pinnatifid, or pinnate ; one-fourth of the rachis is naked, but has numerous black scales."

Observe, also, that the mid-vein of the pinnula is waved ; that the side-veins are forked shortly after leaving the mid-vein ; that, further, the anterior branch of each is elegantly varied, about half-way between the mid-vein and margin, with an elongate, somewhat reniform mass of thecæ, which is partially covered by an indusium attached on the con-cave side of the mass. When approaching maturity, the indusia are forced aside, and ultimately lost, the masses become circular, and often confluent, covering the entire under surface of the pinnula.

This pleasing fern has many names. She is the *Athyrium filix-femina* of Roth and Presl; the *Athyrium irriguum* and *lætum* of Gray; *Asplenium filix-femina* of Bernhardi, Hooker, Mackay, Don, and Francis; the *Polypodium filix-femina* and *rhæticum* of Linnæus, Hudson, and Berkenhout. But however named, and wherever growing,

> " Where the rushing stream is longest,
> There the Lady-fern grows strongest;"

the frequent companion of waterfalls, and mantling many a wild dripping rock with luxuriant vegetation. Seek for the finest specimens, therefore, in places most congenial to their development, and spread them, as previously recommended, between sheets of blotting-paper, where they must remain till perfectly dry.

Nor less pleasing in its rocky habitat is the Spear-shaped Spleenwort, the *Asplenium lanceolatum* of all botanists, the most local of any of the ferns, and generally believed to be confined to the coasts of Merionethshire, Caernarvonshire, Devonshire, and Cornwall, and the neighbourhood of Tunbridge Wells. Those, therefore, who desire to obtain this interesting species, must, in Kent, seek for them on the face of an ivy-mantled cliff near the High Rocks, as also on a similar locality on the Rocks at Tunbridge Wells; in Devonshire, Morwell Rocks, on the banks of the rushing Tamar, are their favourite growing-places, with similar localities beside the Tamar, opposite the Lady Mine; and such as may be seen contiguous to Cann Quarry, on the banks of the Plym. Their long, black, slender, and penetrating rhizoma, which run to a great depth, fix them securely on rugged declivities near St. Ives, and in the Scilly Islands, where, unharmed by tempest, they grow

luxuriantly, and invite many an adventurous botanist to
scale the dizzy heights whereon they grow.

You may not find them in Scotland, nor yet in the
Emerald Isle, whose waters yield the most beauteous sea-

*1st Variety.*        *2nd Variety.*        *3rd Variety.*
SPEAR-SHAPED SPLEENWORT.

weeds.   In Merionethshire they are located near Barmouth,
on rocks and walls; in Caernarvonshire, on a rock to the
left of the road between Tan-y-bwlch and Aberglaslyn, and
on a second rock close to the latter place.

But though the roots (as above mentioned) are long and
penetrating, and evidently designed to anchor the small
fern in its exposed growing-places, it is found occasionally
on old stone walls, sheltered from the winds of heaven.
The rhizoma is brown, tufted, and densely covered with
bristle-like scales ; these should be carefully preserved, as
somewhat of the character of the plant depends on them.
The yong fronds open with the hawthorn and harebells, in
May, when the cuckoo's " one word spoken" rejoices the
young enthusiast who spends his summer holiday in woods
and meadows.   August witnesses their maturity, and
during the months of winter they continue fertile.

Listen now to the distinctive characteristics which pertain to the Spear-shaped Spleenwort, and mark them well; for, otherwise, the differences of the three varieties may elude your vigilance. The form of the frond, as noticed by an experienced botanist, is various, and owes much of its variety to dissimilar situations. "The first is of erect growth, nearly linear, and simply pinnate, the pinnæ being

*1st Variety.*

stalked and lobed. In this form it produces seed abundautly; the masses, when fully grown, are perfectly cironlar; and such is their mode of growth, every part of the frond being perfectly flat, and the entire part ridged."

"A second variety, of pendent growth and larger size, is lanceolate as regards its form; the pinnæ are pinnate; the pinnulæ stalked, serrated, and somewhat quadrate; the

*2nd Variety.*

fronds often reach to a foot in length; they usually issue from dark holes or shaded spots, and the lower pair of pinnæ are often bleached, weak, and of small size—the surface being generally flat, although occasionally slightly concave; when this occurs each pinula partakes more or less of the character, as in the figured specimen, where the detached pinnula shows the veins and incipient indusia."

"A third variety grows nearly erect, but bends over at the extremity; and the entire frond, together with each individual pinnula, possesses such a rigid and inflexible convexity, that it is next to impossible to flatten the plant by pressing it." The form is expressed in the engraving, but the convexity cannot be well described.

The lover of ferns does not readily grow weary while observing the exquisite variety of seeds by which they are distinguished. In the Spear-shaped Spleenwort, the lateral veins are branched, and a branch runs to the extremity of each serrature; the masses of thecæ are affixed near the extremity of the veins, and somewhat alternately, one branch bearing a mass, and the next being without one;

*3rd Variety.*

each mass is at first elongate and linear, and covered by a linear white indusium; the indusium afterwards disappears, and the mass becomes nearly circular.

The Black Spleenwort, *Asplenium adiantum-nigrum* of authors, the *A. lucidum* of Gray, is universally distributed, growing alike in shady places and on rocks open to the sun, though attaining its greatest luxuriance when nestling in the fissures of old walls, amid the rents of ruins, or in damp, shady hedgerows.

BLACK SPLEENWORT.

Engravings may afford accurate sketches of this favourite fern, but they cannot supersede the necessity of minute descriptions, which are indispensable for young botanists, who will, otherwise, be deceived by accidental resemblances to other ferns.

Take notice, therefore, that the root is strong and wiry, and the rhizoma tufted, black, and covered with bristly scales; that the rachis is extremely smooth and shining, having a few scattered scales at the base; and that one-third of the entire length is naked, which portion is usually black, or of a dark purple. You may gather the fronds before the end of May, or in the beginning of June; at first they are nearly erect, but they shortly begin to droop, and finally become quite pendent. September is the season of their maturity, and it is pleasant to look upon them when green and vigorous beneath the leafless branches that often screen them in winter from the wind. Strange contrast are they to denuded oaks and hedgerows, with their dark and cheerless-looking fibres; the rain may fall in torrents, and the fields become white with snow, but still the Black Spleenwort rejoices the passer-by by its ample and luxuriant fronds.

We have spoken of the roots, and the rhizoma, the scales, and rachis. It now remains to notice, that the frond is triangular in form—that the apex is acute and attenuated —that it is pinnate, with triangular pinnæ, acutely pointed, pinnate, and alternate ; whilst the pinnulæ themselves are alternate and triangular, the lower ones pinnate, or pinnatifid, with notched lobes.

It is needful, likewise, to observe that the side-veins in the lobes, or pinnulæ, are irregularly alternate, and mostly forked after diverging from the mid-vein. One or both branches of this divided vein bear an elongate linear mass of thecæ, situated near the mid-vein, and at first completely covered by a long, narrow, white, scale-like indusium, which opens towards the mid-vein. When the thecæ swell, and approach maturity, this small scale is gently raised and pushed from its site ; it is next turned aside, and finally disappears, when the under surface of the frond presents a continuous mass of rich brown seeds.

We may lastly remark, in the words of a brother botanist, " that the superior length of the lower pinnæ, and the oblique angle at which these, and indeed all the pinnæ, are attached to the rachis, with the more central position in the pinnulæ occupied by the thecæ, are characters by which the Black Spleenwort may be readily distinguished."

## OCTOBER.

" Long work it were,
Here to account the endless progeny
Of all the weeds that bud and blossom there ;
But so much as doth need must needs be counted here.''

SPENSER.

WELCOME Nature's handmaids, small mosses, which look green and cheerful when summer's flowers have ceased to deck the meads and banks, and all joyous birds are flown to distant climes ! Ye have laboured much, yet willingly, in the spring-tide of your time, and now ye sit contentedly on the once stern and herbless declivity, mantled by your active ministry with ferns and flowers. I well remember your bleak growing place ; no joyous insect ever settled on its surface, no industrious bee sought there her honey harvest— flowers there were none, neither had the smallest fern struck its tiny roots into the fissures of the rock. At length came one of your persevering sisterhood, making a loving compact with sunbeams and soft showers, with nimble gases and wandering zephyrs ; and when each one had kindly lent his aid, and that small moss, forming to herself a home, was securely settled, others of her companions hastened to dwell beside her. Now, within the fissures, and on rough crags up the stony sides of giant rocks, and nestling on their summits, grow green mosses. Look upon them, passer-by— nay more, gather a few, and carefully examine their component parts. Forest trees do not present a greater or more marked variety ; nor is the stateliest oak or pine, the bread-fruit or cocoa, the banian or the cinnamon, more wondrously adapted to their respective habitats, or more curiously furnished with air and sap vessels for all the functions of vegetable life, than these neglected mosses.

Perchance you have not thought respecting them : to you they are nameless even. Let it not be so, however, for the future ; for every living thing, as I have often said before, hath its own brief history, which we should endeavour to understand—nay more, its characteristic structure and varying embellishments.

Take, for example, the *Bryum triquetrum*, and bear in mind, while looking at it, that the following subdivisions pertain to the *Brya*, which it is needful to remember :—

*Capsules sessile, or nearly so.—Capsules on fruit-stalks, upright.*

| | |
|---|---|
| 1. Stemless. | 3. Stems trailing. |
| 2. Upright. | 4. Stems upright. |

Roundish, egg-shaped, and oblong capsules, pertain to the family of Brya.

*Capsules on fruit-stalks, leaning.*

1. Stem none, or very short and unbranched.
2. Stems upright.

*Capsules on fruit-stalks, drooping.*

1. Stem none, or very short and unbranched.
2. Stems upright.

*B. triquetrum*, or Ventricose bog-fringe moss—assigned by Withering to the family of *Mnium ;* by Messrs Hooker and Taylor to that of *Bryum*—is described by the latter as having branched and elongated stems ; lance-shaped, acute, serrated, and reticulated leaves, with pear-shaped capsules ; the fruit-stalks are very long, and the whole plant answers to the derivation of the generic appellation *Bryum*, derived from a Greek word βρυω, signifying to sprout or shoot up, in allusion to the perpetual greenness and vivifying powers of this somewhat rare plant. Under its name *Mnium*, the Ventricose or bog-fringe moss is mentioned by able botanists as not unfrequent in turf-bogs and marshy places, also

on mud or gravel, by the sides of rivulets and springs, in
the ascent of Snowdon and Glyder Var, or the Hill of Tempests, and on the sandy brink of the river at Mavis Bank,
near Edinburgh. Under its name of *Bryum* the same moss
is, however, noticed by Dr. Scott, in his dried collection, as
being found on the bank of some lake in Ireland, the only
station for this fine plant in the British dominions being thus
vaguely specified by him. The discrepancy is in some
degree accounted for by the admission of Messrs. Hooker
and Taylor, "that the differences between this moss and *B.
mnium cæspiticium* are almost insufficient, and that it is
more distinguishable by its larger size, proliferous habit,
and brown or purple hue, than by any more essential characters." All botanists are aware, that in the same individuals great differences result from soil and station. Such
therefore may be the case with the Bog-fringe moss, respecting which Dellenius mentions that "the red kind is
found in the mountain torrents of Snowdon, the green in
high boggy heaths about London and Oxford." Mr.
Griffith speaks of having gathered it near Celin House, two
miles from Holywell, in Flintshire.

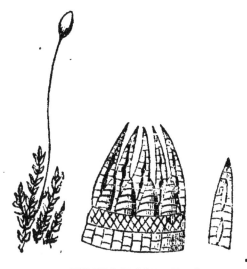

BRYUM TRIQUETRUM.

The peristome of the *B. triquetrum* is compact, and formed with great regularity. Five horn-like projections present two openings in each, and are curiously varied with transverse bands: they are apparently based on four short bee-hive-shaped protuberances, upheld by a circular foundation, with zigzag and banded embellishments.

The peristome of the *B. ventricosum* presents a somewhat different appearance, varied and yet similar. The three sharp pyramidical-shaped processes have a single opening each, while those of *triquetrum* are furnished with two; they have, moreover, intermediate decorations, resem-

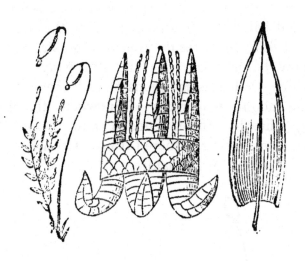

*Peristome.*
BRYUM VENTRICOSUM.

bling strings of upright beads. Observe the broad belt with its seeming scales, and the three unique terminations turned up at the base of each, after the fashion of a Chinese slipper. Ladies might derive from many a wayside moss, when highly magnified, hints for worsted work of no ordinary beauty; those, also, who devise ornamental patterns for artificers of various kinds, might frequently discover elaborate decorations, scrolls, or braiding in many a fringe or capsule that projects from out a bed of moss.

The *B. ventricosum* has also been placed among the *Mniums.* We state, however, on the authority of Messrs. Hooker and Taylor, that this interesting moss pertains to the family of *Bryum*, and these are its characteristics :— Stems elongated, and branched leaves, oblong, acuminated, scarcely serrulate, margins recurved, nerve reaching beyond the point, capsules oblong and pendulous. The stems are frequently from two to four inches or more in height, including innovations. The *ventricosum* delights in marshy ground and the crevices of damp rocks, where it grows abundantly, and is often of a deep brown or reddish hue of which the whole plant generally partakes more or less. The nerve is reddish.

The *B. punctatum*, Dotted fringe moss, is assigned by Withering to the *Mnium* tribe ; by Hooker and Taylor to the *Bryum.* Who may decide between such varying opinions ? We incline, however, to the judgment of the latter authorities, and shall therefore give their concurrent description of this singular moss :—Stems elongated, leaves roundish ovate, very obtuse, reticulated, margins thickened entire, nerves disappearing below the point, capsule ovate, or rather oblong egg-shaped pendulous, lid short, rostrate ; leaves, largest in the order *Musci*, approaching very nearly to those of the *Cinclidium stygium*, inner peristome of a firm and rigid texture, outer teeth pale coloured.

Seek for this interesting moss in marshy places, on the roots of alders and water-loving trees, where it vegetates in large patches, and the leaves have occasionally a scarlet rib. Bogs in the West Riding of Yorkshire are its favourite growing places, and there the brotherhood obtain their fullest development. The broad and inversely egg-shaped leaves are elegantly varied with small dots.

" Prepare your sweetmeats," says the Chinese proverb, "and your friend will come ;" " Think of him, and he will surely appear," responds the English. We trust that the frequent wish for accurate information relative to mosses, will cause this hitherto comparatively neglected branch of natural history to be placed on the same sure basis as sea-weeds and shells. In the instance of the undulated Fringe-moss, or *Byrum undulatum*, the usual difficulty occurs. Withering speaks of it as a *Mnium;* Hedwig also. Hud-

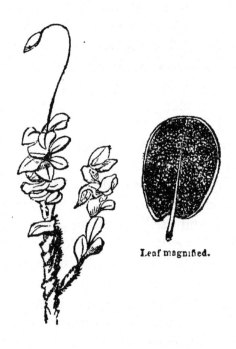

Leaf magnified.

BRYUM PUNCTATUM.

son and Hooker as a *Bryum.* The root is strong and creep-ing ; the shoots from three to six inches long, either branched or otherwise ; the leaves are thin, pellucid, spear-shaped, waved, and serrated ; capsule pendant with blunt lids ; veil straight and pointed, according to Dellenius ; flowers extremely minute, but when examined with a microscope, the unfruitful ones are seen to be surrounded by strap-shaped leaflets, in the centre of shoots ending in mimic roses.

We have elsewhere observed that every plant and flower
has its winged or creeping resident, nay, every locality and
soil: "As for the stork, the fir-trees are her house; the
high hills are a refuge for the wild goats, and the rocks for
the conics." (*Ps.* civ. 17, 18.)

Even the humblest moss that grows on cottage thatch or
wall, in moist shady woods, or about the roots of trees and
hedges, shelters some tiny insect, that finds therein a home
and store-house. On the leaves of this simple moss, and
also upon the *Dicranum bryoides*, or Bryum-like feather-
moss, in Wallington Woods, Northumberland, a minute and
elegant insect, the *Leangeum Trevelyani*, may sometimes
be discovered. None, perhaps, among the insect tribes are
more radiantly beautiful, or more attractive from its
symmetry, delicate formation, and varied hues, than this
minute creature. The eye gazes upon it with extreme delight; and though thus minute, it is endowed with every
requisite faculty for enjoyment; and how know we whether
this charming little being, which presents one of the loveliest objects in creation, does not especially rejoice in the
fairy-formed home wherein she dwells?

Nor less worthy of regard is the *Bryum hygrometricum* of
Huds, the *Furaria hygrometrica* of Roth and Hook, and the
*Mnium hygrometricum* of Withering; for such are the
various appellations assigned to the Revolving fringe-moss.
This tribe is common in woods and heaths, on garden walks,
old trees and walls, among decayed wood, and where coals
and cinders have lain. We remember our favourite walk
in a shrubbery, where the Revolving fringe-moss grew so
luxuriantly as to form a carpet of verdure. Children re-
sorted in summer to that shady walk, and therefore the
green covering was continually trodden upon, although re-
newing its beauty and freshness whenever dews were heavy,
or passing showers watered that wild spot. In places rarely
visited the case was otherwise; there the stems of the Revolving fringe-moss rose from one to two inches in height,

though mostly buried in the earth ; and fruit-stalks, a full
inch long, upheld their pear-shaped, golden yellow capsules.
Those who seek for it in December will discover this same
moss, very small, and nestling closely on the ground, with
fine oblong, taper-pointed leaves, from which young fruit-
stalks project like fairy spears.   In January, the four-sided,
and straw-coloured veil appears ; in February and March,
capsules become apparent, and ripen in April and May.

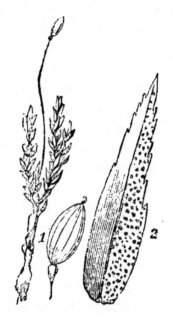

B. PALUSTRE.
1, 2, Capsule and Leaf magnified.

Thus regularly and invariably progresses this small moss;
six months are witnesses to its first emerging and proceeding
through various changes, till its final perfecting by soft
showers and warm sunbeams.   Then it is that the vital
principle is fully developed; and if the fruit-stalk be
moistened at its base, the  head makes three or four revolu-
tions ; but if at the upper portion, it turns the contrary way.
Spiral fibres are, therefore, assigned to the Revolving fringe-

moss—they answer the same end as those of the water-lily, and all such plants as are peculiarly affected by light or moisture.

The essential characteristics of the *Bryum palustre*, or Forked fringe-moss, according to the testimony of Hooker and Taylor, consists in the stems being much branched, the leaves lanceolate, obtuse, entire, with margins revolute; the capsules ovate, oblique, sulcated, lids conical. We recognize in this rarely-noticed species a beautiful provision for the dispersion of those innumerable seeds which are contained in the small ribbed capsules: the stems, previously upright and holding their seed-vessels erect, for the obvious purpose of obtaining as much air and light as possible, bend downwards when the seeds are fully ripe and the lid ready to fall off. When the seeds are shed, the capsules become crooked—wherefore, we know not; yet doubtless for some purpose connected with the vegetable economy of the plant.

Subjects of engrossing interest, either as respects their beauty or rarity, or the memorial sites among which they grow, may be selected from the rugged bank that first attracted our attention. Here is the *Mnium orcuatum*, or Curved-stalked fringe-moss of Dicks and Withering, assigned to the Bryum tribe, an extremely beautiful moss, unknown on the continent, though of common occurrence in the mountainous districts of Ireland, and not unfrequently in many parts of England. The stems are upright, but spreading and serrulated. The barren flowers are terminated and stalk-like; the fruit-stalks terminal, crooked, and surrounded by young shoots. The golden yellow globular capsules have narrow mouths, their fringes are varied with short, upright, acute red teeth, and the minute lids are scarcely beaked. A fuscous woolly substance constantly surrounds the shoots—a material, it may be, for the winter domicile of some dependent insect. Nor less vivid than the capsules are the fruit-stalks, which spring from out the base of the mimic branches; they are golden red, about half-an-inch long, and crooked; the leaves are serrulated chiefly towards the end.

Thus perfect in all its parts, the Curved-stalked fringe-moss is eagerly sought for by collectors; and in order to assist them in their pursuits, we instance the following growing-places:—Bogs in the northern parts of Yorkshire, and moist places on Glyder Mountain, North Wales. Boggy places in Scotland; and among bogs, in company with *Mnium palustre*, in Greenfield, Saddleworth, Yorkshire; and Stanley, Cheshire. The banks of Avon Las, near Pistyllwen, in Llanberris parish, are varied with this elegant moss, as also the foot of the Pentland Hills, near Edinburgh, and the recesses of the mountains of Cumberland.

Beautiful in its assigned locality is the *Bryum ligulatum*, Long-leaved thread-moss, one of the largest as well as handsomest of British mosses. Sir James Smith compares

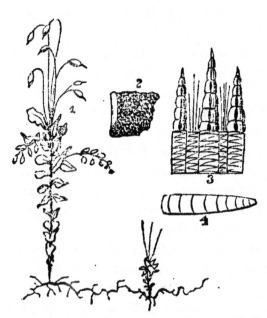

1. B. LIGULATUM.—2. Portion of a leaf, magnified to show the network.—3. Interior of peristome.—4. Exterior.

it to a grove of fairy palm-trees, drooping gracefully over the moon-lit dancers, when they prank it merrily on the dewy sod. Beautiful and shining foot-stalks, resembling polished shafts, of a dark red colour, uphold capsules of equal brilliancy, among which the tiny people may float in and out, now lost in their dark shadows, and again re-appearing in the full beams of a cloudless moon.

Linnæus gave to the Dark Mountain fringe-moss the name which now it bears. Botanists, in modern times, have re-

B. DEALBATUM.   Leaf Magnified.

ferred it to the *Mnium, Dicranum*, and *Trichostomum* tribes; and, therefore, our readers will not be surprised if they find it under either of the above heads. A recent and high authority has, however, resumed the appellation given by the Swedish naturalist.

Unlike many of its family, which grow best on arid moun-tains, and walls open to the sun, the Dark Mountain fringe-moss requires a soil, however meagre. It grows on stones thinly covered with mould, near Llanberris, in Carnarvon-shire, and in the West Riding of Yorkshire; frequent in the Highlands and Lowlands of Scotland; it affects also rocks on the hill-side, about fifty yards above Garthmelio, the seat of R. W. Wynne, Esq., Denbighshire.

A variety, with trailing stems and brownish green leaves, somewhat open, with branched shoots, floating on the water, or mantling stone and pebbles, round which some prattling stream forms eddies of white foam, may be readily distinguished by short and blackish fruit-stalks, and straight, oblong, dark green capsules. This variety is not uncommon in rivulets that water the moors of the Peak of Derbyshire and in the racing torrents near Llanberris, Carnarvonshire. Mr. Griffith gathered a fine specimen in the first brook after crossing Pont Alwen, between Denbigh and Cerrig y Druidion.

The *Bryum dealbatum*, or Pale-leaved thread-moss, is a somewhat rare moss in England, though not unfrequent among Scottish mountains, especially on Ben Lawers. Botanists will readily distinguish it, by observing that the capsules are roundish, somewhat bent, toothed and fringed, and that the leaves are spear-shaped, acute, and expanding. Such are its chief characteristics; and when submitted to a microscope, there is much found to admire in this simple, rock-adhering moss. Minute leaves, which resemble the finest scales, become enlarged, pellucid, finely but obscurely serrated, and seem as if covered with network; the scarcely perceptible capsule displays its teeth and fringes, and delicately formed lid. A pitcher in truth it is, filled with the finest seeds, upheld with others of its kind to the genial influence of air and light, and when that purpose is accomplished, bending to the earth.

The *Bryum marginatum*, or Bordered thread-moss, is equally worthy of attention. The shoots are mostly simple, the leaves egg spear-shaped, pointed, finely toothed, and bordered with a mid-rib and thick red edge, the capsules egg cylindrical, with a beaked lid. All this is obvious to the unassisted eye, and the yellowish hue of the Bordered thread-moss readily distinguishes it; but when seen through a microscope, how greatly is its beauty heightened! The lurid hue of the nerve and margin in each leaf becomes of a brilliant deep blood colour, and the veil is equally observable. True it is that these peculiarities render the moss obvious at first sight, but as the hand of the carver and polisher brings

B. MARGINATUM.—1. Capsule.—2. Portion of a leaf.
3. Leaf magnified.

forth the excellence of cameos and gems, so do the lenses of the microscope alone reveal the exquisite perfections of Nature's minutest works.

## NOVEMBER.

" Mosses are Nature's children.  I have seen them
　　Smile in their beauty on the lone sea-cliff,
　　By rushing torrents, or on herbless granite,
　　Where nought beside, save some meek, pale-faced lichen
　　Would brook to linger."

OTHER individuals of the genus *Bryum* are deserving of minute inspection.  Their numbers and varieties are such, however, as to render selection difficult ; we shall, therefore, briefly notice a few of the rarest, or most beautiful, and then pass on.

The *Bryum argenteum*, or Silvery-thread,moss, though frequent on sunny banks and walls, on roofs and rocks, is deserving of especial notice. The capsules are egg-shaped, upright when green, pendent when ripe; and this for the purpose already noticed, namely, that of scattering the seeds upon the earth as from a reversed pitcher. The leaves are egg-spear-shaped, ending in hairs, but so pressed against the stems, as hardly to be distinguishable by the naked eye; the fruit-stalks rise from the base of the shoots to nearly half an inch; and he who subjects the capsule to a magnifying glass, will readily discover that the lid is short and blunt, that the mouth is elegantly fringed, and the veil deciduous. The plant grows in patches about half an inch high; in autumn, and early in the winter, of a vivid green, then shining and silvery white, especially when dry, a peculiarity which distinguishes the Silvery - thread moss from all others of its brethren.

The gravel-walks of Oxford Physic Garden, in the time of Dillenius, were pleasingly ornamented with variety 2, of which the shoots were pale or dark-green, occasionally shining, the leaves more crowded, and the mouth of the capsule without a fringe. We know not whether this kind still holds its accustomed place; but our botanical friends, who visit Oxford, will do well to seek for it.

*Bryum cubitate*, Elbow-stalked thread-moss, and largest of all the *Brya*, may be readily distinguished by its golden-coloured, reddish-brown, and brightly glittering fruit-stalk, having an elbow-like bend a little above the base, and upholding a depressed and pendant, club-shaped, and very long capsule, with an upright and numerous-toothed fringe. The shoots are somewhat branched, rather re-cumbent at the base, and the stems are trailing, often three inches long, the leaves occasionally bristle-pointed, but not uniformly so.

This interesting species looks well in a moss book. It is agreeably associated with clear, cold streams in the neighbourhood of Snowdon, and with the bonny banks of Aberfeldy. Hooker and Taylor arrange the *Cubitate* under *B. ventricosum;* Griffith considers it as not specifically distinct from *Alpinum.*

Nevertheless, a considerable difference subsists between the *Cubitate* and *Alpinum*, as noticed by Withering; and thus his description runs:—" Densely compact in growth, variously branched, yet irregular. Leaves numerous, oblong, keeled, straight, acute, opaque, smooth, shining, purplish-green; but in old plants, purplish below, dark-red above. Fruit-stalks an inch high, dark-red purple, issuing from a large purple tubercle, veil purplish. This species is best known by its deep shining purple colour, and its rigid stems and leaves: the former remaining perfectly straight even when moistened. Rocks in mountainous regions are the favourite growing places of this beautiful moss, than which few among its brethren look better when carefully dried.

The Great hairy-thread moss, *B. rurale*, friend of the peasant's hut, readily affixes its tiny roots in roofs, whether thatched or tiled, and on walls and the trunks of trees. Linnæus mentions, that when this moss extends over thatched buildings, the thatch, instead of lasting only about ten years, will endure for an age. He suggests, that it may prove a great security against liability to accidents from fire, which renders such covering very objectionable.

Had Linnæus lived in the present age, and seen, as we have lately had frequent occasion to observe, traces of fire among the dry furze and grass which mantle the sides of deep railway cuttings, he might fully have appreciated the value of this moss. Sparks from the fiery iron steed, whirling his living masses of many hundreds of human beings with incredible rapidity, not unfrequently set fire to dry herbage; even to hay-stacks, occasionally, when too near his path; and woe to the peasant's thatched hut that stands beside it!

But he whose roof is covered with this friendly moss, may sleep securely. The snorting of the fiery steed need not disturb his slumbers; his children are safe in bed; flashes of fire—breathings of that tremendous racer—may lighten by his windows, and fall upon his roof; but they do no harm among the dense and elevated tufts of the Great hairy-thread moss. This moss has little of external beauty to commend it except when growing in wide patches, and presenting in its aggregate, during rainy weather, a fine yellowish green, which often pleasingly contrasts with the grey bark of aged trees, or old thatch on barns and cottages. In dry seasons, the same moss looks of a dull grey or brown. As regards its obvious characteristics, we may briefly mention that the capsules are cylindrical, the lids conical and acute, terminated at the mouths with long fringes; that the shoots are branched, the leaves reflected inversely, heart-shaped blunt, hair pointed.

Beautiful specimens for preserving in moss books may be obtained from the family of Thread-moss, both as regards their form and hue. The *B. aureum* is one of these. In this, and in *M. crudum*, the stem is half as long as the fruit-stalk, and extremely shining; the strap-shaped leaves are of a greenish golden hue, forming altogether a firm tuft, and distinguishable by their slenderness and length; the fruit-stalks are an inch and more in length, purple, iridiscent, and issuing from a brownish green involucrum, varying occasionally from pale red to golden yellow, and upholding pear-shaped and green capsules, which, like the supporting shafts, change to yellow red. Although of somewhat rare occurrence, this elegantly varied moss grows on rocks in Nottingham Park, as also among the Berwyn Mountains, in the roads between Bala and Llangunnoy, and on Snowdon.

The transition is natural from mosses to ferns. Companions are they on many a weather-beaten crag, and when the sisterhood of mosses have prepared a dwelling-place for plants of larger growth, ferns are the first to dwell beside them. Such is often the case with the Tunbridge filmy-fern, the *Hymenophyllum Tunbridgense* of Smith and Hooker, of Mackay, Gray, and Francis, the *Trichomanes Tunbridgense*, or Tunbridge golden-locks of Linnæus, Hudson, Withering, Bolton, and Lightfoot.

In England the localities of this singular fern are moist clefts of rocks, and stony places, growing somewhat luxuriantly on the high rocks of Tunbridge Wells; it also embellishes the coast of Sussex, and is found among the pebbles at Cockbush. Many a rushing torrent on Dartmoor reflects its winged leaves; and botanists speak of it as being not unfrequent on the mountains of the north. Mr. Aiken gathered it from crags in a shady dell near Llanberris; Mr. Winch from beside the tumbling falls of the dread Lodore; another botanist at the Cil-hepste Waterfall, near Pont-nedd-vechan, and on Brincous in the vicinity of Neath, Glamorganshire. Variety the 2nd, with fructifications on naked fruit-stalks, is not unfrequent on rocks beneath Dolbaden Castle, near the lake of Llanberris, and on the rock called Foalfoot, on Ingleborough in Yorkshire.

The Tunbridge filmy-fern is presumed to be as yet unknown in Wales and Scotland. But Ireland owns it in various romantic localities, in the counties of Galway and Kerry, Cork and Wicklow. Those who visit the Lakes of Killarney may find it spreading over the rocks in great beauty and luxuriance.

The roots are black, wiry, and slender; the rhizoma creeping, wiry, slender, long, and black. The fronds consist of a branched series of veins, each one being clothed with a membranous or filmy wing; the branches or pinnæ are alternate, more or less subdivided; the subdivisions or pinnulæ are mostly in pairs, the margin of the wing is crenated, and very minutely spiny; the masses of thecæ are in flat marginal receptacles, situated at the union of the

pinnæ with the rachis ; in this species the receptacles have
a serrated external margin.

Hooker mentions the Filmy-fern in his "Flora Lon-
dinensis," as belonging to a very beautiful and extensive

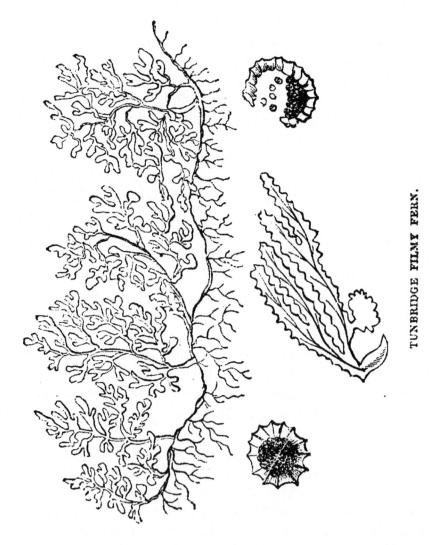

TUNBRIDGE FILMY FERN.

genus, established by Smith, for the most part inhabiting
the tropics. One species alone is European, though not
included in the Floras of Germany and Switzerland,
notwithstanding their rocks and waterfalls, and damp,

shady wood-sides, in which the species congregate. La Bellardiare and Mr. Brown met with luxuriant specimens in New Holland; the former described one especially as a new species, under the name of *Hymenophyllum cupressiforme.*

Hooker speaks also of the *Hymenophyllum alatum*, or Winged-stacked goldilocks, a rare plant, but hitherto imperfectly understood. Ray noticed it in his Synopsis, and also Dillenius, as growing on dripping rocks at Belbank, half-a-mile from Bingley, in Yorkshire, at the well-head of a remarkable spring, and there Dr. Richardson discovered it in modern times. We cannot sufficiently deprecate the ruthless habit of destroying plants by tearing them from their growing places for the sake of preserving only a few specimens. To some such recklessness we owe, most probably, the disappearance of the Winged-stacked goldilocks, or fern, from a spot, consecrated by the visits of Ray and Dillenius, where it grew till the year 1782, and then disappeared, according to the testimony of Hailstone, in Whitaker's "Craven." The species may be readily met with throughout the Snowdon district, and in many parts of the principality of Wales, where harps were heard in unison with the headlong rush of waterfalls; among the Highlands also, in localities too numerous to mention ; and in Ireland, especially at Powerscourt Waterfall, and on shady banks and rocks exposed to the spray of the torrent above Turk Cottage, Killarney, where it grows in company with the rare *Jungermannia Hutchinsiæ.*

The frond is from four inches to four and a half high ; primary pinnæ three inches long, the upper gradually shorter, and these, as well as the secondary ones, are ovate-lanceolate, margin entire, furnished with a slender brown nerve or mid-rib, prominent on both sides, and running down the middle. Rachis winged, with a broad foliaceous margin. The substance of the frond is membranous, smooth, beautifully reticulated, of a brownish-green colour. Capsules roundish sessile, fixed by the disk, compressed brown, collected together near the middle of the receptacle ; the disk in each is reticulated, the elastic ring large, and the seeds round.

Such is the elaborate description given by Hooker, who seems to have regarded the Filmy-fern with no ordinary interest. It was, perhaps, associated in his mind with many a pleasant summer ramble.

The Rue-leaved Spleenwort, *Asplenium rutamuraria* of

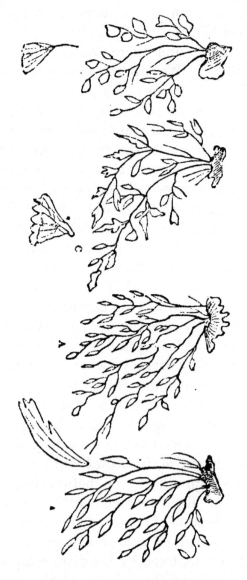

RUE-LEAVED SPLEENWORT

authors, to the generic name of which *Murate germanicum* and *Alternifolium* were assigned by Gray, Willdenow, Wulfen, Smith, and Francis.

Few plants are more generally diffused than the Rue-leaved Spleenwort. Growing abundantly among the rocky hills of Scotland in a perfectly wild state, one of its favourite localities is Arthur's Seat, near Edinburgh, and thither the lover of ferns often hastens to seek for it. One might imagine that the pure breezes, and warm gleams of sunshine, that visit Cader Idris and Snowdon, would favour the growth of this pleasing fern, but such is not the case; travellers who seek for it among those romantic solitudes will find it growing but sparingly. Yet, notwithstanding this restriction, the same fern is common throughout the northern, western, and southern counties of England, as also in Wales, Scotland, and Ireland, where it is found on almost every ruin, old church, or wall, or bridge, whether of brick or stone. The dweller among crowded houses, and the hurry of " street-pacing steeds," who still retains his healthy love of ferns, may readily discover the Rue-leaved spleenwort on the walls of Greenwich Park, though more abundantly rooted in the crumbling mortar that fills interstices in the brick portion of the wall, than in the stone. Like the mouse and sparrow, half-domesticated associates of man, it dwells wherever he has fixed his abode ; associated equally with memorial ruins and the humblest way-side hut, recalling to memory days of feudal splendour, and the peaceful occupations of rural life.

The roots of the *Asplenium ruta muraria* are black and wiry ; the rhizoma is equally black tufted also, and clothed with bristly scales. Associated with the coming back of migratory birds, and the ripening of early fruits, the fronds appear in May and June, arrive at maturity in September, and continue green throughout the winter till the ensuing May. They are invariably fertile. The rachis is black or dark purple, smooth and shining, and for more than half its length, uniformly unclothed. "The normal form of the frond is triangular and pinnate, the pinnæ being alternate, and also pinnate ; the pinnulæ are of varied form, but mostly somewhat triangular or lozenge-shaped ; their exterior margin is generally serrated."

" Veins radiate from the stack to the exterior margin of the pinnulæ, and to these are attached the elongate linear masses of thecæ, two, three, four, and even five on each pinnulæ these are at first covered with an elongate, linear, white indusium, which is pushed aside by the growing thecæ, turned back, and finally lost, the back of the pinnula becoming eventually covered with a dense brown mass of thecæ."

Newman speaks of a remarkable form of the Rue-leaved spleenwort, found in several localities throughout Germany, Hungary, and Scotland, and considered by botanists as a distinct species, under the name of *Asplenium germanicum*, or *A. alternifolium*. A representation of the plant, copied from his *History of British Ferns*, page 71, and named by him the *alternate type* of *A. ruta muraria*, is given above A.A. "The form of the frond is elongate and pinnate ; the pinnæ are distant, small, linear, alternate, and generally notched or divided at the apex. C presents a specimen gathered by Newman, at Arthur's Seat, near Edinburgh. It has three dissimilar fronds, and is introduced as forming a connecting link between the normal type of the common Rue-leaved Spleenwort and the continental specimens. The same botanist remarks that if a naturalist was to commence with the first figured and most frequent, and advance regularly through the others, he would find it difficult to divide the plants into two distinct species. Botanists of eminence, however, have thought otherwise, but the opinion of Linnæus may be quoted as corroborative of the one above mentioned.

Sir S. Smith considers that the Rue-leaved spleenwort is an intermediate species between *Septentrionale* and *Ruta muraria*, though distinct from both.

This singular looking fern, the *Asplenium septentrionale* of Smith, Hooker, Galpine, Gray, and Francis; *Acrostichum septentrionale* of Linnæus, Bolton, Hudson, Lightfoot, Berkenhoot, known by the appropriate name of Forked Spleenwort, is perhaps one of our rarest ferns. It was con-

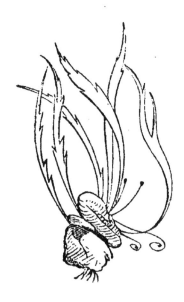

PORKED SPLEENWORT.

sidered for many years as peculiar to Arthur's Seat, but has since been gathered in Carnarvonshire, though sparingly; luxuriantly on a wall by the roadside leading out of Llanrwst, towards Conway, exactly opposite a farm-house, and about a mile from Llanrwst. For the sake of botanists who desire to add this rare plant to their collections, we shall mention its localities.

*England.*—Northumberland, Kyloe Crags, Cumberland, Honiton Crags, and on rocks in the neighbourhood of Scaw Fell.

*Wales.*—Carnarvonshire, Pwll Du, in the romantic Pass of Llanberris; and on Glyder Vawr, above Llyn-y-Coon, very sparingly; a mile from Llanwrst, on the road to

Conway, growing luxuriantly on a wall at the left-hand side.

*Scotland.*—Arthur's Seat and Braid Hills, near Edinburgh, formerly abundant, now rare; Perthshire, near Dunkeld.

*Ireland.*—Unknown.

The roots are long, fibrous, crooked, intertwined, and, together with the rhizoma, which is large and tufted, form an amazing bulk. The specimen procured by Newman, at Llanrwst, had three hundred fronds; and after shaking off a good deal of the earth, the mass of roots and rhizoma weighed several pounds.

The form of the frond is elongate, lanceolate, and furnished laterally with one or two short bifed teeth or serratures; the apex also terminates in a bifed point, diminishing imperceptibly towards the base, and terminating in a smooth rachis, black at the extreme base. The veins are nearly simple, few in number, one uniformly runs into each serrature. The thecæ are attached to each vein in a continuous line, covered at first by an indusium of similar shape, opening towards the mid-vein of the frond, thrown back as the thecæ swell, and finally disappearing. The lower surface of the frond presents a continuous mass of thecæ.

Pleasingly associated with the return of the wandering dove, with the cheerful yet monotonous song of the cuckoo, and the coming back of the swallow family, green leaves of the Forked Spleenwort appear in March and April; they arrive at maturity in August, and retain their verdure throughout winter; they grow in an horizontal position, from out perpendicular or slanting banks or walls, and are figured in their natural size and position.

The Sea Spleenwort, *Asplenium marinum* of authors, is widely diffused throughout England, Scotland, Wales, and Ireland, wherever the fissure of the sea-cliff, or the roof of a marine cave, affords a congenial growing place. Specimens of an enormous size have been gathered from the roof of a large cavern at Petit Bot Bay, in the island of Guernsey, and also in the islands of Madeira and Teneriffe.

Cornwall, with its sea-cliffs, and remembrances of Druidic times, is peculiarly favourable to the full development of the Sea Spleenwort, which grows there to a larger size than in the northern counties.

The root is black and wiry, tough, long, and so firmly fixed in the crevices of rocks, as to require a strong hand for

SEA SPLEENWORT.

its removal. The rhizoma is nearly spherical, black, and covered with bristly scales; the fronds make their appearance in June and July; they ripen their seeds in October, and remain green throughout the year. Fronds of successive seasons may be found equally strong and verdant in July and August.

"The frond is linear, and simply pinnate; the pinnæ are stalked, ovate, and serrated, two larger ones frequently occur near the apex; the pinnæ are connected by a narrow wing running along the rachis."

' Observe how curiously the side-veins are forked almost

immediately after leaving the mid-vein—that the anterior branch bears an elongate linear mass of rust-coloured thecæ; this, when young, is covered by a white membranous indusium of the same shape as the mass, uniformly opening towards the apex of the frond.

Beware, young botanist, while seeking for this fern in marine caverns, or in the fissures of sea-cliffs, that you make yourselves acquainted with the time when the tide rises from day to day. You may otherwise have to wade amid the dashing spray, or else to peril life and limb, in climbing up the slippery sides of rocks, with your hard-earned treasures in your hands; happy, if you escape passing the night on some high crag, which barely affords you a safe standing-place above the strife of waters.

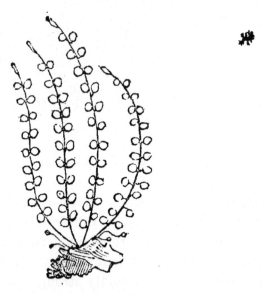

GREEN SPLEENWORT.

The geographical range of the Green Spleenwort, *Asplenium viride* of authors, *A. trichomanes* of Linnæus, is very limited. In Ireland it is believed to be confined to a single mountain —Ben Bulben; in this country to the extreme northern counties, where it is generally intermixed with the *Asplenium trichomanes;* in Scotland, to her loftier mountains; those especially of the western islands; in Wales, to the Snowdon range, though most profusely in the fissure called Twll Dee, and at the base of the fissure where it opens into Cwm Idwell; Cader Idris, Brecon Beacon, the Lady's Waterfall by Neath, and rocks within a few miles of the same place, are also acknowledged localities.

Few among the brotherhood of ferns are more delicately formed than the Green Spleenwort. The root is fibrous, black, and somewhat tender; the rhizoma black, scaly, and tufted. In May and June, when the meadows are bright with flowers, and warbling voices resound from every hedge and thicket, the fronds lift up their heads; they attain maturity in August, and remain green throughout the winter.

For about a third of its length the rachis is uniformly naked; half this part is black er purple; the remainder to the apex of the frond, and all the pinnæ, are of a bright emerald hue; the form of the frond is narrow, elongate, linear, and simply pinnate, and though the pinnæ are not so numerous as in *A. trichomanes,* they are somewhat quadrate, but without angles, and more or less crenate at the margin; they are in general placed alternately on the rachis, are usually very distinct and separate, occasionally crowded, and attached to the rachis by their stalks only.

But the most decided specific character belonging to the plant is, "that the lateral veins are either simple or forked, bearing an elongate linear mass of thecæ, almost immediately on leaving the mid-veins; and that, if forked, the division takes place beyond the mass of thecæ. The veins do not reach the margin of the pinna: the thecæ are at first covered by a linear, elongate indusium, which soon disappears, and they become confluent in a ferruginous mass, occupying the centre of the pinna, and concealing the mid-vein. The masses at first are four or six in number."

## DECEMBER.

" Lonely the forest spring ! a rocky hill
Rises beside it, and an aged yew
Bursts from the rifted crag, that overshades
The waters cavern'd there.  Unseen and slow,
And silently they swell.  The adder's tongue,
Rich with the wrinkles of its glossy green,
Hangs down its long lank leaves, whose wavy dip
Just breaks the tranquil surface."

MANY a pleasant ramble have we taken together, reader,
among the woods and in green lanes, where ferns grew wild
and high, beside the roar of ocean, in quest of such as dwell
on crags, and even in sea-caves, where the marine Spleenwort loves to hide.  Now that trees are leafless, and most of
the fern tribe have retired to their winter quarters beneath
the earth, we must refer to our dried specimens for the four
remaining species which we have still to describe.

Here, then, is the *Scolopendrium vulgare*, the Common
Hart's-tongue, the *S. officinarum*, and *Asplenium scolopendrium*—for such are its three names—a peculiarly handsome
and ornamental fern, which grows alike on streamlet brink
and in the clefts of arid rocks or aged ruins.  Of almost
universal distribution, and with the exception of some parts
of Kent and Northamptonshire, where ferns in general re-
fuse to vegetate, this interesting species is found in every
part of the British empire.  The child who peeps warily
over the edge of a wynch-well, may see its long graceful
leaves reflected in the dark waters beneath ; and those who
rashly peril life or limb in climbing to the sea-crew's nest,
in the slippery sides of wild cliffs, may often notice luxuriant tufts of the same fern waving far above his head.
True it is, that the botanist may walk for miles, and return
disappointed to his home, saying, that nowhere has he been

áble to find the Hart's-tongue ; but let him go over the same ground another day, looking carefully in the thickest parts of hedges, and he may return with fully-developed specimens. In Scotland, where the *Scolopendrium* is

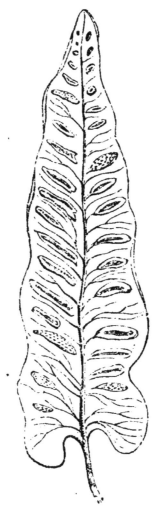

HART'S-TONGUE.

sparingly distributed, he who searches can seldom travel far without recognising it in some favourite locality. In Ireland it is far more abundant, and is not only profusely scattered in the most dissimilar situations, but attains almost

giant growth in the neighbourhood of Sligo, and among the
romantic solitudes of Killarney, where the fronds, radiating
from a common centre, arch gracefully in a semicircle.

The roots may be briefly described as black, of considerable length and thickness, and of great tenacity; the
rhizoma is tufted, scaly, blackish, and almost spherical.
Simultaneously with the arrival of the cuckoo, and the
flowering of the cowslip and marsh-marigold in meadows,
the Hart's-tongue uplifts its head, often in their immediate
vicinity, and pleasingly contrasting its light green leaves
with the delicate yellow petals of the one, and the brighter
tints of the other. Though storms are abroad, and snows
lie deep upon the ground, we might find this hardy species
in its sheltered haunts; for the fronds, which arrive at full
maturity by the end of September, continue green and
vigorous throughout the winter, and generally await the
springing forth of a fresh progeny in April.

DISSIMILAR LEAVES OF THE HART'S-TONGUE.

The habit of the plant is well marked, and is decidedly
different from every other species. Any further description
is therefore needless, except to notice that the form of the
frond is elongate, linear, and undivided; acute at the apex
or termination, and cordate at the base. Such is the case
when fully developed; but seedling plants present a variety
of forms, and the young botanist will do well to remember
the peculiarities which they occasionally assume.

Those who like to collect memorial plants, may find dwarf specimens on the old gateway leading to Saltwood Castle, in Kent—last halting-place of the murderer of Thomas á-Becket; luxuriant ones, on the left hand bank leading from that once stronghold of feudal splendour, where they grow in company with several other species.

It is more than probable that the plant mentioned by Gerard under the name of *Hemionitis sterilis*, found by him in a gravelly lane leading to Oxey Park, near Watford, fifteen miles from London, and also on the wall of Hampton Court, was no other than the Hart's-tongue. " It is a very small and base herb," wrote he, " not above a finger high,

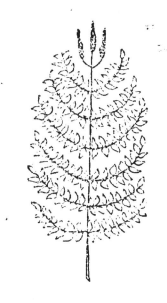

OSMUND ROYAL.

having four or five small leaves of the same substance and colour, and spotted on the back like unto Hart's-tongue." The dwarfish appearance which the old herbalist describes, may be ascribed without doubt to its sterile growing-place : for, although individuals of the species grow luxuriantly in the clefts, or on the summits of high rocks, this peculiarity

results from the moisture imparted by clouds and vapours
in their lofty domiciles. We have frequently had occasion
to notice, that plants which prefer humid situations in low
grounds, thrive equally well on hill-tops, and that for the
same reason.

Hail to the banks of Loch Tyne, and those of far-famed
Killarney! Our specimen was gathered in the first home
of the Flowering-fern, or French brachen—the *Osmunda
regalis* of authors—the crown prince of English ferns.
Though widely diffused throughout various portions of Great
Britain and Ireland, the species is nowhere more abundant
or luxurious than in the above-mentioned localities, rising
at one time to the height of eight feet, at another bending
gracefully over the water's edge. This peculiarity is very
obvious at Killarney, where the long fronds form arcades of
verdure, affording a welcome shelter to the nimble coot,
from whence she gazes fearlessly on the tourist, though
often skimming near in his rapid boat. Beautiful are the
lakes, and mountains, and trees of this wild spot; and yet
Sir Walter Scott, when visiting Killarney, utteied not a
word in praise of them, till he reached the place where
grew the Flowering-fern, and then it was that he broke
silence, saying, " *This* is worth coming to see." " And
truly," wrote Newman, to whom we are indebted for the
anecdote, " I did not wonder at the great man's taste; to
me it appeared the most wonderfully beautiful spot I had
ever beheld, and this beauty is mainly owing to the immense
size and number of " the French brachen's pendant fronds."

Widely is this fern distributed, and yet its " metropolis"
appears to be the west of Ireland, more particularly Connemara, where it not unfrequently covers the smaller islands
with a carpet of verdure; those in the centre being generally
rigid and erect, such as grow around the margin pendulous.
You cannot mistake it wherever growing, as, with the exception of the lonely Moonwort, no other fern bears its seeds
in spikes. The roots are strong and fibrous; the rhizoma

tufted, and very large, and hence capable of annually producing such a weight of foliage; young fronds, varying in
number from six to twelve, appear in May, and attain maturity in August. Unlike the Hart's-tongue, they cannot
bear the severity of winter, but shrink from the first frosts,
and presently disappear. No sooner, however, do the
sullen clouds of an ungenial season pass away, and fierce
winds cease to howl through forest walks, than they come
forth from their hiding-places with a rapid and vigorous
growth, and, until nearly brown, present a reddish hue.
The fronds are fertile and barren.

We owe to Dr. Withering the appropriate name of
"flower-crowned prince of English ferns," by which he
designates the lordly brachen. He speaks of it as affording
a curious instance in its seeds of long-suspended vitality,
as the plant, though previously unknown for many miles
around Birmingham, suddenly appeared on an archery-
butt at Moseley Common, artificially raised with mud from
a deep pit, wherein the seeds had probably lain for a great
length of time. Do not fail to procure this interesting
species; it is very available for rock-work, especially if
removed with a portion of bog-earth, and can scarcely fail
to produce an ornamental effect wherever growing. Take
care to avoid cutting with the spade its enormous rhizoma;
when this is done the plant becomes so much weakened
as scarcely to recover its pristine vigour; but, should the
injury accidently occur, observe that the rhizoma has a
whitish core or centre, termed by old Gerard, in his
" Herbal," " the heart of Osmund the Waterman."

Botanists trace in the Moonwort—*Botrychium lunaria* of
Smith and Hooker, the *Osmunda lunaria* of Linnæus, which
often, from its diminutive size, escapes that notice which
the stately *Osmunda regalis* cannot elude—a fancied resemblance to the moon, presented by its leaves, and which
has caused it to be held in superstitious reverence. Many a
youth and damsel have gone forth beneath the clear calm

beams of the full moon, to gather its "leaves of power,"
printing the dewy sod with noiseless tread, and dreading to

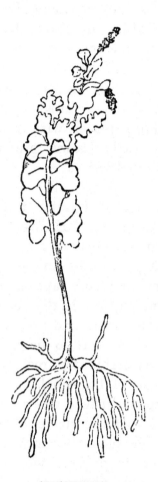

MOONWORT.

look around.   And still the cottage by the wood remains,
as poets tell, with its bee-hives and gushing streams—

> " Whence rapidly, with foot as light
> As young musk roe's, out she flew,
> To cull each shining leaf that grew,
> Beneath the moonlight's hallowing beams."

Singular varieties occasionally occur, but the specimen in
our fern-book is the most frequent.   Though widely dis-

tributed in various parts, it is yet somewhat rare, and is more widely diffused in England than in either Wales, Scotland, or Ireland. The root differs materially from that of the true fern, as also the rhizoma, which appears little more than a subterraneous portion of the root. Newman, whose observations on this favourite branch of natural history are derived from personal inspection, notices that before the Moon-wort has felt the influence of spring it exists in a quiescent state, consisting of a simple stem scarcely an inch in length, and placed vertically in the earth, somewhat attenuated at the lower extremity, while the upper has a whitish bud-like termination, the embryo frond of the coming season. That part of the rhizoma which especially derives nourishment from the earth, bears two distinct whorls of thick yellowish succulent roots; the upper portion is encased in alternate scale-like sheaths, and the elongation of the rhizoma shows that the young frond is about to shoot. The frond, which is almost universally a single one, appears in April or May, erect and straight, as if to welcome the coming back of the swallow people! It is fertile except in seeding plants, and occasionally reaches the height of six inches.

The Adder's-tongue (*Ophioglossum vulgatum*) is generally distributed throughout England, but is comparatively scarce in Wales, Scotland, and Ireland. Its favourite growing-places are moist, damp meadows, and the sides of streamlets, where the scarlet Lychnis loves to nestle; and is occasionally so abundant as to cover acres of grass-land with its long, smooth, hollow frond, appearing in May, and withering at the latter end of August. A few only of the fronds are fertile, and from out the acute and slanting, the deep green and leafy portion of such, uprises a straight, erect, club-shaped spike, somewhat longer than the leafy part, and bearing seeds in a double longitudinal row. When the seeds are fully ripe, you may readily see the gradual opening of the thecæ transversely, waiting, as it were, for the passing-by of autumn winds, that bear them in their airy chariots over dale and hill.

Thus ends our pleasant converse about the family of British Ferns and Mosses. As regards the first, we have transferred, from their growing-places to the leaves of our Herbarium, all known species; and while remembering the many healthy walks which we have taken together, let us not forget their names and characteristic peculiarities: and when the voice of Spring again summons the sleeping tribes, let us go forth to welcome them. As respects the second, we have briefly noticed a very few; trusting that the beauties or peculiarities of such may incline the votaries of nature to desire a farther acquaintance with their tribe.

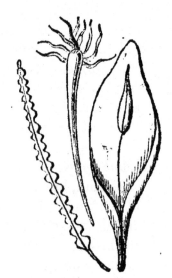

ADDER'S-TONGUE.

London: Printed by H. Teck, New Street, Cloth Fair.